Robust Filtering and Fault Detection for T-S Fuzzy Systems

Xiao-Lei Wang · Guang-Hong Yang ·
Yu-Long Wang

Robust Filtering and Fault Detection for T-S Fuzzy Systems

Springer

Xiao-Lei Wang
School of Mechatronic Engineering
and Automation
Shanghai University
Shanghai, China

Guang-Hong Yang
College of Information Science
and Engineering
Northeastern University
Shenyang, Liaoning, China

Yu-Long Wang
School of Mechatronic Engineering
and Automation
Shanghai University
Shanghai, China

ISBN 978-981-96-5820-6   ISBN 978-981-96-5818-3 (eBook)
https://doi.org/10.1007/978-981-96-5818-3

This Springer imprint is published by the registered company Springer Nature Singapore Pte Ltd.
The registered company address is: 152 Beach Road, #21-01/04 Gateway East, Singapore 189721, Singapore

Preface

Modern control systems exhibit multiple characteristics such as high complexity, uncertainty, large time-delay, and strong coupling, etc. A large number of configured actuators and sensors are prone to generate faults in complex and ever-changing working environments, leading to a deterioration in the performance of the control systems. Therefore, it is required that the control systems can generate alarms timely and accurately in the case of faults, and take corresponding measures to complete the established tasks. Fault detection technology is an effective method to solve this problem. In addition, due to the widespread presence of uncertainty, nonlinearity, and external factors in actual dynamic systems, it is difficult to establish mathematical models that satisfy the control requirements, and therefore, satisfactory control results cannot be achieved through the use of traditional control theory. The proposal and development of Takagi-Sugeno (T-S) fuzzy model provides an effective way to solve this problem. However, it should be noted that the dynamics and stability of the systems are difficult to obtain in T-S fuzzy modeling, which requires studying and dealing with the performance issues of the systems regarding uncertain factors. Robust filtering of the control systems is an effective technical solution to solve this problem. At present, significant progress has been made in the research of robust filtering and fault detection for T-S fuzzy systems. However, due to the inherent nonlinear characteristics of T-S fuzzy models, many issues need to be further studied. For example, the introduction of event-triggered mechanisms will lead to the mismatch of the premise variables, making it difficult to design fuzzy filters and fault detection observers. Therefore, it is still an important research topic to explore more effective robust filtering methods and fault detection methods for T-S fuzzy models.

Structure and Readership

This book conducts in-depth research on robust filtering and fault detection for a class of T-S fuzzy systems, consisting of eight chapters. Chapter 1 is an introduction, which introduces the research background and significance of this book, and elaborates on the current research status of robust filtering and fault detection for T-S fuzzy systems at home and abroad, and provides preliminary knowledge, relevant theories, lemmas, and mathematical symbols involved in this book. In Chap. 2, piecewise fuzzy filters are established at different instants for discrete-time T-S fuzzy systems with immeasurable state variables, and a fuzzy filtering scheme is designed to solve the problem of the existing methods being unable to deal with the external disturbances, while achieving better results. In Chap. 3, an improved delays-dependent region partitioning method is proposed for discrete-time T-S fuzzy systems with multiple time-varying delays, and a fuzzy filter design scheme is established to achieve better robust performance than the existing results. Chapter 4 studies the problem of robust filtering for continuous-time T-S fuzzy systems with bounded external disturbances via premise-region-dependent event-triggered mechanisms, and the purpose of designing the fuzzy filter and the event-triggered mechanism separately is realized. Chapters 5 and 6 focus on the design of event-triggered fault detection observers for T-S fuzzy systems, and an improved matching membership function method and online asynchronous premise reconstruction method are, respectively, proposed to solve the asynchronous problems of premise variables and overcome the defects of the existing methods. In addition, in Chap. 7, a new fault detection observer design solution to address the false alarm caused by measurement outliers in fault detection systems is investigated. In Chap. 8, conclusions and prospects are discussed.

Shanghai, China
Shenyang, China
Shanghai, China

Xiao-Lei Wang
xiaoleiwang@shu.edu.cn

Guang-Hong Yang
yangguanghong@ise.neu.edu.cn

Yu-Long Wang
yulongwang@shu.edu.cn

Acknowledgments

The content of this book is a summary of the author's phased work in T-S fuzzy systems, robust filtering, and fault detection over the past seven years, as well as an exploration of the existing technical methods. The research work was completed under the careful guidance of the second author, Prof. Guang-Hong Yang from Northeastern University, who is the supervisor of the first and third authors. Professor Yang's profound academic achievements, rigorous research style, and selfless work enthusiasm have left a deep impression on us, and are also our lifelong model. Here, we would like to express our high respect and heartfelt gratitude to Prof. Yang! In addition, we would like to express the gratitude to Jasmine Dou from Springer-Verlag and the Series Editor for their valuable support.

The research work involved in this book was supported in part by the Funds of the National Natural Science Foundation of China (Grant Nos. 62203082, 52371372, and U23A20337).

Contents

Chapter 1
Introduction

The main content of this chapter is to introduce the background knowledge of the research, analyze the existing results, and provide theoretical support for the subsequent work. The contents of this chapter are summarized as follows: First, the research background of Takagi-Sugeno (T-S) fuzzy system are briefly discussed. Then, the research status of robust filtering and fault detection is summarized. Finally, the main research content of this book is briefly described.

1.1 Research Background

With the development of science and technology, automatic control systems have been applied in various fields such as industrial enterprise production, aerospace control and monitoring, and have also promoted tremendous progress in modern society. Unlike the singularity and simplicity of traditional linear control systems, modern control systems exhibit multiple characteristics such as high complexity, uncertainty, large time-delay, and strong coupling, etc., it is difficult to establish mathematical models that satisfy the control requirements, resulting in many control theory methods being unable to be applied in practical engineering. The proposal and development of T-S fuzzy model provides an effective way to solve this problem. In addition, it is difficult to obtain accurate information of the system states due to the high operating cost of the actual systems. At the same time, the systems are prone to the influence of unknown external noise and delay signal, which will deteriorate the system performance and even destroy the stability of the systems. Therefore, it is of great theoretical value and practical significance to realize the normal operation of the systems and maintain certain performance even after the uncertainty, external interference, time-delay and unknown system states appear in the actual systems.

X.-L. Wang et al., *Robust Filtering and Fault Detection for T-S Fuzzy Systems*,
https://doi.org/10.1007/978-981-96-5818-3_1

In addition, the actual systems are highly complex, and maintain long-term, high-power, high-load operation, which will lead to the sensor mechanism, the actuator mechanism and the internal components of the systems are highly susceptible to changes in the internal and external environment of the systems, resulting in inevitable varying degrees of aging, and even faults. The existence of faults in the system will inevitably lead to a deterioration of system performance, and even destroy the stability of the systems, resulting in economic losses, as well as catastrophic consequences such as personal injury and death. For example, in 2017, the Long March 5 Yao II rocket took off from the Wenchang Space Launch Site. At 346 s of flight, the engine suffered a local fault in a complex force-thermal environment, resulting in a sharp decrease in engine thrust and ultimately failed the launch mission. Afterwards, the relevant personnel invested a lot of time in troubleshooting the faults, and successfully carried out the second launch on the basis of fully implementing the fault improvement measures. In fact, the faults that cause such accidents can be promptly identified through pre-designed fault diagnosis mechanisms, thereby avoiding the failure of launch missions.

In recent years, information technology and network communication technology have greatly promoted the development of networked control systems, and the amount of data transmitted through the network is also increasing. In traditional time-triggered methods, data is sampled with a fixed period and then updated for transmission, which will lead to the transmission of unnecessary data and further waste of network resources. In addition, due to the limited network bandwidth in actual network environments, transmitting unnecessary data will also cause network congestion. To address the above issues, a new communication mechanism, namely, event-triggered mechanism, was proposed, which has attracted widespread attention from scholars due to its ability to perform non-periodic sampling, save network resources, and improve resource utilization. The issues of event-triggered for different types of networked systems have been studied and fruitful results have been achieved. However, it should be noted that the current research on the event-triggered problem of T-S fuzzy systems is still immature and needs to be further explored.

1.2 Overview of T-S Fuzzy Systems

In the1960s scholar L.A. Zadch published the achievement "Fuzzy Sets" in Ref. [1], which proposed the concept of fuzzy systems and introduced the foundation of language models, proposing the definition of fuzzy logic, thus laying the foundation for the basic theory of fuzzy control.

Fuzzy systems are composed of three types of systems, including pure fuzzy systems, fuzzy systems with fuzzifiers and defuzzifiers, and T-S fuzzy systems. A pure fuzzy system consists of three parts, namely the fuzzy rule library, fuzzy inference machine, and the relationship between fuzzy set spaces. The advantage of this system is that it facilitates the mapping expression of multiple inputs and outputs, while the disadvantage is that the inputs and outputs are all fuzzy sets. However, the input and output data are required to be accurate values in practical systems, thus the application range of pure fuzzy systems is limited. A fuzzy system with a fuzzier and a defuzzifier consists of three parts, namely a fuzzy rule library, a fuzzy inference machine, and an input-output relationship between the fuzzy combination and the defuzzifier. The advantage of this system is that the input and output data are truth-value variables, which has a wide range of applications. The disadvantage is that it requires more reliance on expert experience knowledge in controller design, and the stability of the control systems cannot be studied theoretically. In order to overcome the shortcomings of the two methods mentioned above, the T-S fuzzy model was proposed by Takagi and Sugeno in Ref. [2] in 1985, which provides a theoretical basis for the transformation of knowledge base to nonlinear mappings and provides theoretical support for the analysis and synthesis of the practical systems.

The T-S fuzzy model can be used to describe the dynamic characteristics of a class of complex nonlinear systems, and is composed of the premise parts and the consequent parts. The qualitative knowledge described by fuzzy rules is called the premise parts, and the quantitative knowledge described by local linear systems is called the consequent parts [3–5]. In addition, the premise part and consequent part are established as a global fuzzy model through the superposition of membership functions [6–8]. It has been demonstrated in Refs. [9–12] that nonlinear functions can be approximated by T-S models on a closed compact set with arbitrary precision. In this way, classical linear system theory and fuzzy system theory are used to deal with the theoretical problems of nonlinear systems. Therefore, T-S fuzzy theory provides a theoretical basis for dealing with the problems related to nonlinear systems.

In the past few decades, numerous scholars have conducted research on T-S fuzzy theory, and have proposed a large number of controller and filter design methods for T-S fuzzy models [14–20]. For example, in Refs. [10–13], the state feedback control problem and the output feedback control problem are respectively studied for a class of systems based on Lyapunov stability theory. In Refs. [21, 22], a fault detection scheme was constructed to successfully detect actuator faults in T-S fuzzy systems by establishing appropriate Lyapunov functions. In addition, under the framework of Lyapunov theory, the robust filtering problem of T-S fuzzy systems was studied in Refs. [23, 24]. However, it should be noted that the research results mentioned above are all constructed based on the quadratic Lyapunov function methods, which has been proved to be conservative with the deepening of the study of Lyapunov functions. To address this issue, a series of new types of Lyapunov functions have been proposed by scholars, including fuzzy Lyapunov functions [25–28], polynomial Lyapunov functions [29, 30], piecewise Lyapunov functions [31–33], and a class of non-monotonic Lyapunov functions [34, 35]. Although the research on T-S fuzzy systems has achieved fruitful results, many problems need to be further studied and improved due to the inherent nonlinear characteristics of T-S fuzzy models.

1.3 Current Status of Robust Filtering

Robust filtering and state estimation are two important problems in control theory research and practical engineering applications. The interference of unknown factors to the actual systems will cause the actual measurements to be affected by the noise, which will lead to the deterioration of system performance or even instability of the control systems. Therefore, it is necessary to study and address the performance issues of the system related to uncertain factors. In addition, some information required for system analysis and synthesis often exists in the system states, and cannot be obtained due to technical limitations and high costs. In this case, it is necessary to estimate or reconstruct these unknown signals, and the robust filtering of control systems provides an effective way to solve the above problems.

1.3.1 Overview of Robust Filtering

In the past few decades, since the optimal filtering theory and observer theory were proposed by Kalman and Luenberger, researchers have conducted extensive research on the filtering problem of control systems. The Kalman filtering theory is widely applied in random processes in practical engineering, where precise mathematical models are required and the statistical information of random noise is known [36–39]. However, the external noise in actual systems is uncertain, and the statistical information of random noise cannot be obtained. Therefore, the application of Kalman filtering theory has certain limitations [40, 41]. Therefore, it is of great theoretical significance and practical value to study the robust filtering for the systems with uncertainty and non-Gaussian noise. Based on the above reasons, the robust filtering theory has been proposed and has been favored by many scholars.

In 1981, the robust H_∞ filtering theory based on norm minimization was proposed by scholar G. Zames, which is considered to be the beginning of the development of modern robust control theory. In early H_∞ filtering theory, the H_∞ theory was directly transformed into the H_∞ algorithm assuming that the upper bound of system measurement noise was known. Subsequently, many scholars proposed a series of robust filtering methods based on different practical needs and the energy characteristics of uncertain signals, which can be divided into the following categories:

(1) Robust guaranteed cost filtering
 Similar to the problem of robust guaranteed cost control, the basic idea of robust guaranteed cost filtering is to design filters for the system, and then require stability conditions to be satisfied by the filtering error system, while also requiring a certain degree of guaranteed cost performance to be satisfied. Refer to Refs. [42, 43].
(2) Robust H_∞ filtering
 External uncertainty and disturbances need to satisfy the corresponding requirements in robust H_∞ filtering, that is, the square of the external perturbation signal

is required to be energy bounded. The basic idea is to require the H_∞-norm of the transfer function of the filtering error system to be less than a predetermined bound, which is essentially to minimize the influence of the disturbance signal on the error system. The H_∞ robust filter has become the most widely used one since it was proposed by Grimble and Elsayed (1989), and a large number of robust H_∞ filtering problems have been studied with the introduction of linear matrix inequality methods, and fruitful research results have emerged. Refer to [44, 45].

(3) Robust H_2/H_∞ filtering
The requirement for external disturbances in robust H_2/H_∞ filtering is that some of the statistical information of measurement noise is known, and the other part is required to be a signal with bounded energy. The purpose of robust H_2/H_∞ filtering is to ensure that the variance of the estimation error satisfies a certain performance index, while also requiring the instantaneous value of the estimation error to satisfy the H_2 and H_∞ performance within a certain range. Refer to [46, 47].

(4) Robust L_1 filtering
External disturbance signals that exist in practical systems typically have persistent characteristics, such as sinusoidal input signals and step signals, etc., in which these signals are amplitude bounded rather than energy bounded. Therefore, the H_∞ method for such signals will no longer be applicable, and instead, L_1-gain filtering will be chosen to characterize the robustness, where the L_1 filtering is applicable to the scale of peak describing the signal, i.e. peak to peak filtering. Refer to [48, 49].

(5) Robust L_2-L_∞ filtering
The signals that satisfy both energy bounded and amplitude bounded exist in practical systems, and in this case, the existing robust filtering methods will no longer be applicable. Robust L_2-L_∞ filtering is proposed to deal with the case where the input signals are energy bounded and the output signals are amplitude bounded. Therefore, the robust L_2-L_∞ filtering is also named energy-peak filtering. Refer to [50, 51].

1.3.2 Overview of Robust Filtering for T-S Fuzzy Systems

At present, the robust filtering methods for linear systems have been developed relatively mature, while designing filters for nonlinear systems is still relatively difficult. The combination of T-S fuzzy model and linear matrix inequality method was proposed earlier in Ref. [52] to study the filter design problem of nonlinear systems. However, it should be pointed out that system uncertainty was not considered in the system in Ref. [52]. In the following decades, research on the filtering problem of T-S fuzzy systems has achieved fruitful results [53–56], and some representative results will be discussed. The design of fuzzy filters for T-S fuzzy systems with non-uniformly sampled was studied in Ref. [57]. A non-fragile distributed H_∞ filter design condition was established for discrete-time T-S fuzzy systems in sensor networks using the robust control method in Ref. [58]. By constructing a Lyapunov functional with a factor of α, an exponential H_∞ filter was designed for switched delay systems in Ref. [59]. A general sensor fault model was established for the T-S fuzzy model, and the design conditions for a reliable H_∞ filter were constructed in Ref. [60].

Note that in the results discussed above, the premise variables of the fuzzy filter are consistent with those of the fuzzy model, which implies that all the premise variables in the fuzzy model can be accessed online. However, the premise variables in many T-S fuzzy models are not always obtainable online, for example, the premise variables in the T-S fuzzy models depend on the unmeasurable states. Generally speaking, the methods for constructing T-S fuzzy models can be summarized into two categories, namely identification from input-output data and direct derivation from given nonlinear system equations [61]. The premise variables in the T-S fuzzy model constructed by using the second method often depend on the states of the system, and thus, the premise variables that depend on the unmeasurable states cannot be accessed. The methods proposed above are not applicable to T-S fuzzy systems with unmeasurable premise variables. Therefore, in the past decade, various robust filtering methods have been proposed by scholars for such systems. By incorporating the weights of unreachable fuzzy rules into uncertainty, the filtering and control design is transformed into an H_∞ filtering and control problem for uncertain fuzzy systems in Ref. [62]. In Ref. [63], the robust filtering problem of fuzzy systems was solved by using the partially measurable premise variables of the fuzzy model as the ones of the fuzzy filter.

At present, although some achievements have been made in the research of robust filtering for T-S fuzzy systems, there are still many shortcomings, and new filter design methods need to be explored to overcome the limitations and conservatism of the existing methods.

1.4 Current Status of Fault Detection

Fault diagnosis technology is extremely important in practical engineering and is widely applied in fields such as aviation, aerospace, navigation, and intelligent manufacturing, etc., where the fault refers to one or more observable or calculable variables or characteristics in the control system deviated from the normal value range [64]. The complexity and difference of modern control systems lead to a diversity in the location and time of faults. Thus, in order to avoid faults caused by environmental factors and sudden changes in operating conditions, which can damage the machine and cause economic losses, it is necessary to respond to the faults as much as possible and prevent their malicious effects. Therefore, it is of great significance to improve the reliability and safety of system operation by studying the fault detection of practical systems.

1.4.1 Overview of Fault Diagnosis

The existing fault detection technologies can be divided into two categories, namely data-based fault detection methods and model-based fault detection methods. Among them, data-based methods mainly rely on historical data, such as principal component analysis methods and partial least squares methods [65]. Model-based fault detection methods are developed on the basis of hardware redundancy methods, which, as the name suggests, refer to the use of multiple hardware to repeatedly complete the measurement of the same signal [66]. For example, three sensors are used to measure the outputs of the same systems. The analytical redundancy method is called model-based fault detection method, which implies that the mathematical models and related estimation methods are used to complete the fault detection. Compared with the hardware redundancy methods, the redundant hardware is not needed in analytic redundancy methods, thus it has absolute advantages in cost saving, and has attracted the attention of researchers in recent years. Generally speaking, fault diagnosis using analytic redundancy method can be divided into three steps. First, a fault detection observer is designed for the systems to construct residuals that are robust to disturbances and system uncertainties, while also being sensitive to faults. Then, the fault detection logic unit is established to complete the fault detection and isolation. Finally, a fault isolation observer is further designed to judge the size and shape of the faults.

In practical engineering, faults can be summarized into three types based on their occurrence time, which are:

(1) Abrupt fault: Abrupt fault refers to fault that occurs suddenly in a short period of time in a system, usually caused by sudden changes in internal parameters or components of the system.
(2) Slow variation fault: Slow variation fault refers to the fault that slowly occurs and changes within the system.
(3) Intermittent fault: Intermittent fault refers to the fault caused by internal aging or environmental factors that result in poor contact of internal components in the systems.

Faults can be summarized into three types based on the location of the faults, which are:

(1) Actuator fault: Actuator fault refers to the occurrence of a fault in the actuator, resulting in the actuator not being able to correctly execute the control instructions of the system, further causing an impermissible deviation between the actual inputs and actual outputs of the actuator.
(2) Sensor fault: Sensor fault refers to the fault occurring in the sensor unit of the control system, resulting in some or all of the system information being unable to be accurately measured, further causing an impermissible deviation between the measurement outputs and the actual outputs of the system.
(3) Component fault: Component fault refers to the mutation of certain components in the system, resulting in abnormal operation of the system.

Fault diagnosis theory is a comprehensive research field, including robust theory, adaptive theory and artificial intelligence theory, which has become an important branch of the control field in the past decades and plays a crucial role in modern control systems. Therefore, it is of great significance to carry out the research on fault diagnosis theory. Fault diagnosis generally considered in the research of fault diagnosis theory includes three aspects, which are:

(1) Fault detection: The purpose of fault detection is to determine whether a fault has occurred in the system.
(2) Fault isolation: The purpose of fault isolation is to determine the location and type of faults in the system.
(3) Fault estimation: The purpose of fault estimation is to determine the size and shape of the faults in the system.

Fault detection is the most fundamental component of fault diagnosis, with the purpose of determining whether a fault has occurred in the system, and the fault detection device will issue an alarm in the case of a fault in the system. Fault isolation is an intermediate stage of fault diagnosis, and the location and type of the fault need to be judged when the detection device issues an alarm. Fault estimation is the final stage of fault diagnosis, and more fault information, such as the size and shape of the faults, will be obtained in this stage. In addition, in the practical application of fault diagnosis, according to the actual requirements of three fault diagnosis performance indicators need to be considered, respectively:

(1) Sensitivity performance: The ability of the actual system to detect faults is defined as fault sensitivity, in general, the higher the sensitivity, the more sensitive the system to faults, and the stronger the detection ability of the system.
(2) Robust performance: The ability of the system to suppress model uncertainty and external interference is defined as robust performance. In fault detection problems, it is required that the system can still detect the occurrence of faults in the presence of model uncertainty and external disturbance. Generally speaking, the stronger the robustness, the better the fault detection performance.
(3) Fault detection rate: The fault detection rate includes the false positives and the false negatives, where the false positives refers to the occurrence of a fault in the system but the alarm device has no alarm, and the false negatives refers to the absence of a fault in the system but the alarm device has issued an alarm.

With the increasing complexity of modern engineering systems, fault diagnosis technology is playing an increasingly important role in ensuring the reliability and safety of the system. To ensure the safe production of engineering systems and prevent the occurrence of catastrophic accidents, it is necessary to find and deal with faults promptly and accurately. Therefore, the research on fault diagnosis of control systems has important theoretical value and practical significance. In the past few decades, research on fault diagnosis methods for linear systems has achieved fruitful results and established a relatively complete theoretical system [67–69]. However, there are still many problems that need to be addressed in fault diagnosis for nonlinear systems.

1.4.2 Overview of Fault Detection for T-S Fuzzy Systems

Due to the accurate approximation ability of T-S fuzzy models to nonlinear systems in local regions, fault detection for T-S fuzzy systems has received widespread attention from scholars [70, 71]. At the same time, the T-S fuzzy system is composed of locally linear systems connected by applying the IF-THEN rule, thus the linear system control theory is also applicable to the fault detection of T-S fuzzy systems. In the past few decades, scholars have conducted in-depth research on fault detection of T-S fuzzy models, mainly using robust fault detection methods, which require fault detection systems to be robust to model uncertainties, external disturbances, and other uncertain factors, while also being sensitive to faults. For T-S fuzzy systems, the main fault detection methods are summarized as follows:

(1) H_∞ observer based fault detection method. The specific idea of this method is to design a fault detection observer for T-S fuzzy system to generate residual signals, and realize the sensitivity of residual signals to faults and the robustness to external disturbances and internal uncertainties, such that the fault diagnosis problem is transformed into a convex optimization problem presented in the form of linear matrix inequalities. In Refs. [72, 73], the fault detection problem of a class of T-S fuzzy systems was studied by using an H_∞ observer-based fault detection method.
(2) Sliding mode observer based fault detection method. In this method, fault detection is achieved by utilizing a sliding mode variable structure observer that is insensitive to parameter changes and disturbances, without the need for online system identification, and therefore, it is easy to implement and can achieve fast response to faults. In recent years, the research based on this method has achieved fruitful results and formed some excellent theoretical schemes for fault detection [74–76].
(3) Adaptive observer based fault detection method. In this method, the system performance can be dynamically adjusted based on the input signal, thereby achieving online parameter estimation. Moreover, due to its simple design, strong adaptability, ability to asymptotically estimate the upper bound of faults, and excellent estimation performance for slow variation faults, a large number of researches on fault detection have been carried out by using this method [77–79].
(4) State augmentation based fault detection method. In this method, the fault signals are defined as new state variables and augmented with the original system states to build a new system. Then, synchronous estimation of the system states and fault signals is achieved by designing a fault detection observer for the augmented system. The results of applying this method to fault diagnosis of T-S fuzzy models are presented in Refs. [80–82].

In addition, with the development of digital technology, achieving remote fault detection of equipments has become a research hotspot in the field of fault diagnosis, and is widely used in unmanned detection in military and remote device diagnosis in industry. The connection between the system model and the fault detection observer is point-to-point in traditional fault detection, and the outputs of the system are sampled and then directly transmitted to the fault detection observer. Therefore, the transmission of a large amount of data in the limited network bandwidth will inevitably lead to the blockage of network resources, and the transmission of unnecessary data will also cause the waste of network resources. In order to solve the above problems, a new communication mechanism, named event-triggered mechanism, has been proposed, which adopts on-demand transmission instead of traditional point-to-point transmission, saving network resources while limiting the loss of system performance. Therefore, the research on event-triggered fault detection for T-S fuzzy systems has become a focus of research in recent years and has achieved fruitful results [83–86]. However, it should be noted that the introduction of event-triggered mechanisms for T-S fuzzy system will cause that some measurement information, including premise variables, cannot be accurately received by the fault detection observer, resulting in a mismatch between the premise variables in T-S fuzzy system and the ones in fault detection observer, thus increasing the difficulty in the formulation of fault detection schemes. In order to overcome this problem, several methods have been proposed by researchers, including the non parallel distributed compensation (non-PDC) method proposed in Refs. [21, 83] to deal with the asynchronous problem of premise variables. The main idea of this method is that the function of the system states and the function of the fault detection observer states are selected as the premise variables in system model and the ones in observer, respectively. Moreover, the membership functions in observer and the ones in system can also be inconsistent, and the mismatching of the premise variables can be solved by establishing inequalities to constrain the membership functions. Although this method has some flexibility in the selection of membership functions, it actually has conservatism. In Ref. [86], a partitioning method is introduced to deal with the asynchronous problem of premise variables. The space region of premise variables is divided into the clear regions and the fuzzy regions, and the regions where the premise variables in system and the ones in fuzzy observer are located at different instants are considered, so as to solve the mismatch of the premise variables. Although this method can effectively handle the asynchronous problem of premise variables, the piecewise Lyapunov functions cannot be used in continuous-time T-S fuzzy models, which has the limitation of application. New asynchronous premise reconstruction method is proposed in Refs. [87, 88] to handle the asynchronous problem of premise variables. However, it is important to note that the proposed methods proposed require the upper and lower bounds of λ to be given before the controllers are designed. However, it is known from (13) to (17) in Refs. [87, 88] that the upper and lower bounds of λ cannot be computed before the controllers are calculated. In this case, the results in Refs. [87, 88] will not be able to perfectly solve the mismatch of premise variables.

To sum up, the current research on fault detection for T-S fuzzy systems has achieved fruitful results, and conservatism and limitations still exist in many research schemes, and therefore, it is necessary to explore an effective asynchronous solution.

1.5 Preparatory Knowledge

In order to facilitate the description and understanding of this book, the commonly used symbols, T-S fuzzy system theory and several important lemmas are presented in this chapter.

1.5.1 Symbol Description

- Z^n represents the n-dimensional Euclidean apace, S^T represents the transpose of the matrix S, A^{-1} represents the inverse of matrix A.
- $diag\{\cdot\}$ represents the diagonal matrix, where "$\cdot$" is defined as the diagonal element in the diagonal matrix.
- For a symmetric matrix M, $M > 0$ and $M < 0$ denotes positive definiteness and negative definiteness, respectively.
- The symmetric terms in a symmetric matrix are denoted by $\star$.
- $\lambda_{\max}(A)$ is the largest eigenvalue of A, $\lambda_{\min}(A)$ is the smallest eigenvalue of A.
- I_n represents the identity matrix with n-dimensional, 0 represents the zero matrix with a appropriate dimension.
- $\|\cdot\|$ denotes the Euclidean norm of the vector "$\cdot$".
- $\text{sgn}(\cdot)$ represents a standard symbolic function.
- Define $X(t)$ as a square-integrable signal, and $\|X(t)\|_2 = (\int_0^\infty \|X(t)\|_2 dt)^{\frac{1}{2}}$, and then, $\|X(t)\|_2$ can be expressed as the energy of $X(t)$. The total of all finite energy signals is defined as L_2 and given as follows:

$$L_2 = \{X(t) : \int_0^\infty \|X(t)\|_2 dt < \infty\}$$

  where $\|X(t)\|_2$ is named the L_2-norm of $X(t)$.
- Define $f(t)$ as a amplitude bounded signal, and $\|f(t)\|_\infty = \sup_{t\geq 0} \|f(t)\|$. When $\|f(t)\|$ is a scalar signal, $\|f(t)\|_\infty$ represents the peak of $f(t)$. Define the total of all amplitude bounded signals as L_∞, as follows:

$$L_\infty = \{f(t) : \|f(t)\| < \infty\}$$

  where $\|f(t)\|_\infty$ is named the L_∞-norm of $f(t)$.

1.5.2 T-S Fuzzy Theory

The T-S fuzzy model was proposed by Takagi and Sugeno [2] in 1985 and is used to describe the dynamic characteristics of a class of complex nonlinear systems. The T-S fuzzy model is composed of the premise parts and the consequent parts, in which the qualitative knowledge described by fuzzy rules is named the premise parts, and the quantitative knowledge described by local linear systems is named the consequent parts. In addition, the premise parts and consequent parts are established as a global fuzzy model through the superposition of membership functions. The details are shown as follows.

Plant Rule $\mathcal{E}^i$: If $g_1(x(k))$ is ϵ_{i1}, $g_2(x(k))$ is ϵ_{i2}, ..., $g_p(x(k))$ is ϵ_{ip}, then

$$x(k+1) = A_i x(k) + D_i w(k)$$

where $g(x(k)) = \left[g_1(x(k)) \; g_2(x(k)) \; \cdots \; g_p(x(k)) \right]$ represents the vector of the premise variables, $i \in \mathcal{E} \triangleq \{1, 2, \ldots, r\}$ is the number of inference rules, ϵ_{ik} represents the fuzzy set, $x(k) \in R^n$ an $w(k) \in R^w$ represent the system state and external disturbance, respectively. A_i and D_i are known constant matrices with appropriate dimensions.

Then, by using the fuzzy inference method with a singleton fuzzifier, product inference, and a center-average defuzzifier, the dynamic global model is obtained as:

$$x(k+1) = \frac{\sum_{i=1}^{r} v_i(g(x(k)))(A_i x(k) + D_i w(k))}{\sum_{i=1}^{r} v_i(g(x(k)))}$$

in which

$$\sum_{i=1}^{r} v_i(g(x(k))) = \prod_{j=1}^{n} \epsilon_{ij} g_j(x(k))$$

$\varsigma_{ij}(g_j(x(k)))$ is the grade of membership of $g_j(x(k))$ in ς_{ij}, and it is assumed that:

$$v_i(g(x(k))) \geq 0, \sum_{i=1}^{r} v_i(g(x(k))) > 0$$

Denoting

$$h_i(g(x(k))) = \frac{v_i(g(x(k)))}{\sum_{i=1}^{r} v_i(g(x(k)))}$$

and

$$0 \le h_i(g(x(k))) \le 1, \sum_{i=1}^{r} h_i(g(x(k))) = 1$$

Then, the global T-S fuzzy model is established, as shown below:

$$x(k+1) = \sum_{i=1}^{r} h_i(g(x(k)))(A_i x(k) + D_i w(k)).$$

1.5.3 Several Lemmas

Lemma 1.1 ([31]) (S-procedure)

(1) For $Z_1(Y) = Y^T Q_1 Y + 2X_1^T Y + R_1$, assuming the existence of matrix $\tilde{Y} \in R^m$ such that inequality $Z_1(\tilde{Y}) > 0$ holds, then the following inequalities are equivalent:

1. For all $Y \in R^m$ that satisfy $Z_1(Y) \ge 0$, $Z_0(Y) = Y^T Q_0 Y + 2X_0^T Y + R_0$ can be obtained;
2. The existence of $\lambda \ge 0$ makes the following inequality feasible:

$$\begin{bmatrix} Q_0 & X_0 \\ X_0^T & R_0 \end{bmatrix} + \lambda \begin{bmatrix} Q_1 & X_1 \\ X_1^T & R_1 \end{bmatrix} \ge 0$$

(2) For $Z_1(Y) = Y^T Q_1 Y \ge 0$, assuming that the existence of $\tilde{Y} \in R^m$ satisfies $Z_1(\tilde{Y}) > 0$, the following conditions are satisfied:

1. For all $Y \in R^m$ that satisfy $Z_1(Y) \ge 0$, $Y^T Q_0 Y > 0$ holds;
2. The existence of $\lambda \ge 0$ leads to $Q_0 - \lambda Q_1 > 0$ being satisfied;

Lemma 1.2 ([33]) (Schur completement lemma) For symmetric matrix $X = \begin{bmatrix} X_{11} & X_{12} \\ X_{12}^T & X_{22} \end{bmatrix}$, where X_{11} is $m \times m$ dimensional, the following three conditions are equivalent:

(1) $X < 0$.
(2) $X_{11} < 0$, $X_{22} - X_{12}^T X_{11}^{-1} X_{12} < 0$.
(3) $X_{22} < 0$, $X_{11} - X_{12} X_{22}^{-1} X_{12}^T < 0$.

Lemma 1.3 ([89]) The following conditions are equivalent:

(1) There is a symmetric matrix $P > 0$ such that:

$$A^T PA + Q < 0$$

(2) There are symmetric matrices $P > 0$ and G such that

$$\begin{bmatrix} Q & -A^T G^T \\ -GA & P - G - G^T \end{bmatrix}.$$

1.6 Publication Outline

The existing research on robust filtering and fault detection for T-S fuzzy systems has been introduced in the previous section. It can be seen that in the past few decades, related issues have always been a hot topic of research and have achieved fruitful results, and some of the results have also been successfully applied to practical engineering. However, it should also be noted that many shortcomings still exist in the existing technical solutions. First, in order to achieve better filtering results, the existing filter design schemes cannot effectively solve the robust filtering problem of T-S fuzzy system under the condition of constructing the complex Lyapunov function. Then, the co-design of filter and event-triggered mechanism is necessary in the existing research results of event-triggered robust filtering, which leads to high computational complexity and time-consuming. Subsequently, in order to solve the problem of event-triggered fault detection for T-S fuzzy systems, various methods for handling the asynchronous problem of premise variables have been proposed, all of which have a certain degree of conservatism and limitations, and there is still a lack of an effective mechanism to deal with the asynchronous problem. Finally, the fault detection and alarm device in the actual system is usually affected by the measurement outliers, resulting in false alarms. Although a wealth of fault detection methods have been proposed over the past few decades, research on the false alarm problem is still in its infancy.

In this book, new design solutions are proposed for T-S fuzzy systems to address the existing problems in the research of robust filtering and fault detection. The book consists of eight chapters, with the main content as follows.

Chapter 1: The research background of this book is introduced in this chapter, and the research status of robust filtering and fault detection of T-S fuzzy systems is described. Moreover, the preliminary knowledge is presented in this chapter, and the relevant theories, lemmas, and mathematical symbols involved in this book are introduced.

Chapter 2: Piecewise fuzzy H_∞ filter design for T-S fuzzy systems. This chapter is concerned with the design of H_∞ filters for T-S fuzzy systems. By decomposing the space of premise variables into crisp regions and fuzzy regions, piecewise fuzzy H_∞ filters are established at different instants to cope with the challenges caused by the mismatch of the premise variables and exogenous disturbances. Based on the proposed new Lyapunov functions that depend on the different regions where the premise variables are located, sufficient conditions in terms of linear matrix inequalities (LMIs) are obtained to guarantee the stability and desired performance of the filtering error systems. Compared with the existing design methods, the technical solution proposed in this chapter can achieve better results.

Chapter 3: H_∞ filtering for T-S fuzzy systems with multiple time-varying delays. This chapter is concerned with the design of H_∞ filters for discrete-time T-S fuzzy systems with multiple time-varying delays. An improved delays-dependent region partitioning method is proposed to reduce the conservatism of the existing results. and a new Lyapunov function is established based on this method to prove that the filtering error system can achieve the desired performance. Moreover, the results obtained with an improved delays-dependent region partitioning method proposed in this chapter can be reduced to the existing ones.

Chapter 4: Event-triggered filtering for T-S fuzzy systems with bounded disturbances. This chapter studies the problem of robust filtering for continuous-time T-S fuzzy systems with bounded external disturbances. First, attention is focused on the analysis and design of piecewise fuzzy filters without considering the event-triggered mechanisms. Then, based on the established piecewise fuzzy filter, event-triggered piecewise fuzzy filter with time regularization is designed, such that the new filtering error system is Zeno-free and L_∞ stable, in which new event-triggered conditions that depend on the different regions where the premise variables of the T-S fuzzy model and the piecewise fuzzy filter are located are designed. Unlike the existing results, the method proposed in this chapter can achieve separate design of fuzzy filters and event-triggered mechanisms.

Chapter 5: Event-triggered fault detection observer design for T-S fuzzy systems. This chapter investigates the design of event-triggered fault detection observers for T-S fuzzy systems. An improved matching membership function method is proposed, and the equality constraints are established under the proposed technical framework, such that the membership functions of the residual generator can be obtained directly without calculation. Then, a fault detection scheme is established to ensure the desired performance of the fault detection system. It is shown that the proposed method can overcome the shortcomings of the existing results.

Chapter 6: Event-triggered switched fault detection observer design for T-S fuzzy systems. This chapter is concerned with the problem of event-triggered fault detection observer design for T-S fuzzy systems. New types of observers, namely, switched fuzzy fault detection observers, are designed, which are switched at different triggering instants based on the online asynchronous premise reconstruction method. By employing new triggering instants-dependent Lyapunov functions established via the idea of event interval partitioning, it is proved that the proposed fault detection observers can guarantee the stability and desired performance of the systems. Compared with the existing fault detection methods, the fault detection scheme proposed in this chapter can achieve better fault detection results.

Chapter 7: Fuzzy fault detection observer design for T-S fuzzy systems with measurement Outliers. This chapter is concerned with the problem of fault detection for T-S fuzzy systems with measurement outliers. A new fuzzy fault detection observer with adaptive saturation of output errors is first constructed to reduce the effect of measurement outliers on the error systems and the residual signals, thereby avoiding the occurrence of false alarms. The fault detection scheme is established to achieve fault detection. Compared with the existing fault detection results, the proposed fault detection scheme can effectively avoid false alarms caused by measurement outliers.

Chapter 8: The conclusions of this book are presented in this chapter, and the prospects are also discussed.

References

1. Zadeh LA (1965) Fuzzy sets. Inform Control 8(3):338–353
2. Takagi T, Sugeno M (1985) Fuzzy identification of systems and its applications to modeling and control. IEEE Trans Syst Man Cybern 1:116–132
3. Wu Y, Dong JX (2019) Local stabilization for discrete-time T-S fuzzy time-delay systems with sensor fault. Fuzzy Sets Syst 374(1):115–137
4. Wei YL, Qiu JB, Shi P, Lam HK (2017) A new design of H-infinity piecewise filtering for discrete-time nonlinear time-varying delay systems via T-S fuzzy affine models. IEEE Trans Syst Man Cybern: Syst 47(8):2034–2047
5. Zhao T, Dian SY (2017) Fuzzy dynamic output feedback H_∞ control for continuous-time T-S fuzzy systems under imperfect premise matching. ISA Trans 70:248–259
6. Cao SG, Rees NW, Feng G (1997) Analysis and design for a class of complex control systems part II: Fuzzy controller design. Automatica 33(6):1029–1039
7. Wang LX, Mendel JM (1992) Fuzzy basis functions, universal approximation, and orthogonal least-squares learning. IEEE Trans Neural Netw 3(5):807–814
8. Sala A, Guerra TM, Babuska R (2005) Perspectives of fuzzy systems and control. Fuzzy Sets Syst 156(3):432–444
9. Li WF, Xie ZC, Zhao J, Chu SQ, Wong PK, Gao JW (2022) Improved AET robust control for networked T-S fuzzy systems with asynchronous constraints. IEEE Trans Cybern 52(3):1465–1478 (2022)
10. Feng G (2006) A survey on analysis and design of model-based fuzzy control systems. IEEE Trans Fuzzy Syst 14(5):676–697
11. Lam HK, Narimani M (2009) Stability analysis and performance design for fuzzy-model-based control system under imperfect premise matching. IEEE Trans Fuzzy Syst 17(4):949–961
12. Lam HK (2018) A review on stability analysis of continuous-time fuzzy-model-based control systems: from membership-function-independent to membership-function-dependent analysis. Eng Appl Artif Intell 67:390–408
13. Peng C, Ma SD, Xie XP (2017) Observer-based non-PDC control for networked T-S fuzzy systems with an event-triggered communication. IEEE Trans Cybern 47(8):2279–2287
14. Yang FW, Han QL, Liu YR (2019) Distributed H_∞ state estimation over a filtering network with time-varying and switching topology and partial information exchange. IEEE Trans Cybern 49(3):870–882
15. Wang YL, Lim CC, Shi P (2017) Adaptively adjusted event-triggering mechanism on fault detection for networked control systems. IEEE Trans Cybern 47(8):2299–2311
16. Yang KH (2015) Hierarchical control of a class of uncertain linear systems. J Control Decis 3(2):119–131

17. Zhang JH, Feng G (2014) Event-driven observer-based output feedback control for linear systems. Automatica 55(7):1852–1859
18. Wu LG, Gao YB, Liu JX, Li HY (2017) Event-triggered sliding mode control of stochastic systems via output feedback. Automatica 82:79–92
19. Li XJ, Yang GH (2016) FLS-based adaptive synchronization control of complex dynamical networks with nonlinear couplings and state-dependent uncertainties. IEEE Trans Cybern 46(1):171–180
20. Tong SC, Sui S, Li YM (2015) Fuzzy adaptive output feedback control of MIMO nonlinear systems with partial tracking errors constrained. IEEE Trans Fuzzy Syst 23(4):729–742
21. Pan YN, Yang GH (2018) Event-triggered fault detection filter design for nonlinear networked systems. IEEE Trans Syst Man Cybern 48(11):1851–1862
22. Dong HL, Wang ZD, Lam J, Gao HJ (2012) Fuzzy-model-based robust fault detection with stochastic mixed time delays and successive packet dropouts. IEEE Trans Syst Man Cybern, Part B (Cybern) 42(2):365–376
23. Su XJ, Wu LG, Shi P (2013) Sensor networks with random link failures: distributed filtering for T-S fuzzy systems. IEEE Trans Indus Inform 9(3):1739–1750
24. Tao J, Lu RQ, Su HY, Shi P, Wu ZG (2018) Asynchronous filtering of nonlinear Markov jump systems with randomly occurred quantization via T-S fuzzy models. IEEE Trans Fuzzy Syst 26(4):1866–1877
25. Yang DS, Xie XP (2017) Observer-based state estimation of discrete-time nonlinear systems via a novel maximum-priority-based fuzzy observer. Signal Process 137:63–68
26. Wang LK, Lam HK (2019) New stability criterion for continuous-time Takagi-Sugeno fuzzy systems with time-varying delay. IEEE Trans Cybern 49:1551–1556
27. Yang DS, Xie XP (2016) Relaxed H_∞ control design of discrete-time Takagi-Sugeno fuzzy systems: A multi-samples approach. Neurocomputing 171(1):106–112
28. Wang LK, Liu XD, Zhang HG (2018) Further studies on H_∞ observer design for continuous-time Takagi-Sugeno fuzzy model. Inform Sci 422:396–407
29. Ye D, Diao NN, Zhao XG (2018) Fault-tolerant controller design for general polynomial-fuzzy-model-based systems. IEEE Trans Fuzzy Syst 26(2):1046–1051
30. Lam HK, Seneviratne LD (2011) Stability analysis of polynomial fuzzy-model-based control systems under perfect/imperfect premise matching. IET Control Theory Appl 5(15):1689–1697
31. Wang HM, Yang GH (2016) Fault detection and isolation for affine fuzzy systems with sensor faults. IEEE Trans Fuzzy Syst 24(5):1058–1071
32. Peng S, Su XJ, Li F (2016) Dissipativity-based filtering for fuzzy switched systems with stochastic perturbation. IEEE Trans Autom Control 61(6):1694–1699
33. Qiu JB, Feng G, Gao HJ (2010) Fuzzy-model-based piecewise H_∞ static-output-feedback controller design for networked nonlinear systems. IEEE Trans Fuzzy Syst 18(5):919–934
34. Ahmadi AA, Parrilo PA (2008) Non-monotonic Lyapunov functions for stability of discrete time nonlinear and switched systems. In: 2008 47th IEEE conference on decision and control, pp 614–621
35. Derakhshan SF, Fatehi A, Sharabiany MG (2014) Nonmonotonic observer-based fuzzy controller designs for discrete time T-S fuzzy systems via LMI. IEEE Trans Cybern 44(12):2557–2567
36. Chui CK, Chen G (2008) Kalman filtering with real-time application, 4th edn. Springer, Berlin
37. Julier SJ, LaViola JJ (2007) On Kalman filtering with nonlinear equality constraints. IEEE Trans Signal Process 55(6):2774–2784
38. Plarre K, Bullowon F (2009) On Kalman filtering for detectable systems with intermittent observations. IEEE Trans Autom Control 54(2):386–390
39. Sinopoli B, Schenato L, Franceschetti M, Poolla K, Jordan MI, Sastry SS (2004) Kalman filtering with intermittent observations. IEEE Trans Autom Control 49(9):1453–1464
40. Gershon E, Limebeer D, Shaked U, Yaesh I (2001) Robust H_∞ filtering of Stationary continuous-time linear systems with stochastic uncertainties. IEEE Trans Autom Control 46(11):1788–1793

41. He Y, Wang Q, Lin C (2006) An improved H_∞ filter design for systems with time-varying interval delay. IEEE Trans Circ Syst II: Express Briefs 53(11):1235–1239
42. Kim J, Oh H, Park H (2001) Guaranteed cost and H_∞ filtering for time-delay systems. In: Proceeding of the American control conference, pp 4014–4018 (2001)
43. Petersen IR, Mcfarlanbe DC (1994) Optimal guaranteed cost control and filtering ceratin linear systems. IEEE Trans Autom Control 39:1971–1977
44. Dong JX, Yang GH (2009) H_∞ controller synthesis via switched PDC scheme for discrete-time T-S fuzzy systems. IEEE Trans Fuzzy Syst 17(3):544–555
45. Wang YC, Wang R, Xie XP, Zhang HG (2018) Observer-based H_∞ fuzzy control for modified repetitive control systems. Neurocomputing 286(19):141–149
46. Jose GC (1999) Optimal linear filtering parameter uneertainty. IEEE Trans Signal Process 47:168–175
47. Chen BS, Lee HC, Wu CF (2015) Pareto optimal filter design for nonlinear stochastic fuzzy systems via multiobjective H_2/H_∞ optimization. IEEE Trans Fuzzy Syst 23(2):387–399
48. Kim W, Won S (2000) Robust indeed l_∞-norm control for uncertain discrete-time systems: an LMI approach. IEICE Trans Fundam 83:558–562
49. Adedor J, Nagpal K, Poolla K (1996) A linear matrix inequality approach to peak-to-peak gain minimization. Int J Robust Nonlinear Control 6:899–927
50. Gao HJ, Wang CH (2003) Delay-dependent robust H_∞ and L_2-L_∞ filtering for a class of uncertain nonlinear time-delay systems. IEEE Trans Autom Control 48:1661–1666
51. Choi HD, Ahn CK, Karimi HR, Lim MT (2017) Filtering of discrete-time switched neural networks ensuring exponential dissipative and l_2-l_∞ performances. IEEE Trans Cybern 47(10):3195–3207
52. Feng G, Lee TH, Zhang N (1998) Stable filter design of fuzzy dynamic systems. IEEE Int Conf Fuzzy Syst 1:475–480
53. Wang XL, Yang GH (2019) Event-triggered H_∞ filtering for discrete-time T-S fuzzy systems via network delay optimization technique. IEEE Trans Syst Man Cybern: Syst 49(10):2026–2035
54. Dong J X, Yang G H. Dynamic output feedback H_∞ control synthesis for discrete-time T-S fuzzy systems via switching fuzzy controllers. IEEE Trans Syst Man Cybern, Part B (Cybern) 38(4):1166–1175
55. Chang XH, Yang GH (2014) Non-fragile H_∞ filter design for discrete-time fuzzy systems with multiplicative gain variations. Inform Sci 266:171–185
56. Liu JX, Wu CW, Wang ZH, Wu LG (2017) Reliable filter design for sensor networks using type-2 fuzzy framework. IEEE Trans Indus Inform 13(4):1742–1752
57. Zhang H, Shi Y, Wang J (2014) On energy-to-peak filtering for nonuniformly sampled nonlinear systems: a Markovian jump system approach. IEEE Trans Fuzzy Syst 22(1):212–222
58. Zhang D, Cai W, Xie L, Wang QG (2015) Nonfragile distributed filtering for T-S fuzzy systems in sensor networks. IEEE Trans Fuzzy Syst 23(5):1883–1890
59. Wang D, Shi P, Wang J, Wang W (2012) Delay-dependent exponential H_∞ filtering for discrete-time switched delay systems. Int J Robust Nonlinear Control 22(13):1522–1536
60. Tian E, Yue D (2013) Reliable H_∞ filter design for T-S fuzzy model based networked control systems with random sensor failure. Int J Robust Nonlinear Control 23(1):15–32
61. Tanaka K, Wang HO (2001) Fuzzy control systems design and analysis: A linear matrix inequality approach. Wiley, New York, NY, USA
62. Yoneyama J (2008) H_∞ output feedback control for fuzzy systems with immeasurable premise variables: discrete-time case. Appl Soft Comput 8(2):949–958
63. Dong JX, Yang GH (2017) H_∞ filtering for continuous-time T-S fuzzy systems with partly immeasurable premise variables. IEEE Trans Syst Man Cybern: Syst 47(8):1931–1940
64. Himmelblau DM (1978) Fault detection and diagnosis in chemical and petrochemical process. Elsevier Press, Amsterdam, Holland
65. Ding SX (2014) Data-driven design of fault diagnosis and fault-tolerant control systems. Springer, London

66. Beard RV (1971) Failure accommodation in linear system through self reorganization. Masschusetts Instituteof Technology, MA, USA
67. Koenig D (2005) Unknown input proportional multiple-integral observer design for linear descriptor systems: application to state and fault estimation. IEEE Trans Autom Control 50(2):212–217
68. Huang J, Wang Y, Fukuda T (2018) Set-membership-based fault detection and isolation for robotic assembly of electrical connectors. IEEE Trans Autom Sci Eng 15(1):160–171
69. Ma YK, Tang WT, Fei ZY, Sun XM (2024) Dynamic event-triggered sensor fault interval estimation and accommodation for aeroengine. IEEE Trans Indus Electron 71(4):4191–4200
70. Li FB, Shi P, Lim CC, Wu LG (2018) Fault detection filtering for nonhomogeneous Markovian jump systems via fuzzy approach. IEEE Trans Fuzzy Syst 26(1):131–141
71. Yang GH, Wang HM (2015) Fault detection and isolation for a class of uncertain state-feedback fuzzy control systems. IEEE Trans Fuzzy Syst 23(1):139–151
72. Yang H, Xia Y, Liu B (2011) Fault detection for T-S fuzzy discrete systems in finite frequency domain. IEEE Trans Syst Man Cybern, Part B-Cybern 41(4):911–920
73. Zhao Y, Lam J, Gao HJ (2009) Fault detection for fuzzy systems with intermittent measurements. IEEE Trans Fuzzy Syst 17(2):398–410
74. Castillo-Toledo B, Anzurez-Marin J (2005) Model-based fault diagnosis using sliding mode observers to Takagi-Sugeno fuzzy model. In: Proceedings of the 2005 IEEE international symposium on intelligent control, Limassol, Cyprus, pp 652–657
75. Akhenak A, Chadli M, Ragot J, Maquin D (2008) Fault detection and isolation using sliding mode observer for uncertain Takagi-Sugeno fuzzy model. In: 16th Mediterranean conference on Control and Automation Congress Centre, Ajaccio, France, pp 286–291
76. Shen Q, Jiang B, Cocquempot V (2012) Fault-tolerant control for T-S fuzzy systems with application to near-space hypersonic vehicle with actuator faults. IEEE Trans Fuzzy Syst 20(4):652–665
77. Zhang K, Jiang B, Shi P (2012) Fault estimation observer design for discrete-time takagisugeno fuzzy systems based on piecewise Lyapunov functions. IEEE Trans Fuzzy Syst 20(1):192–200
78. Jiang B, Gao Z, Shi P, Xu Y (2010) Adaptive fault-tolerant tracking control of near-space vehicle using Takagi-Sugeno fuzzy models. IEEE Trans Fuzzy Syst 18(5):1000–1007
79. Qi R, Tao G, Jiang B, Tan C (2012) Adaptive control schemes for discrete-time T-S fuzzy systems with unknown parameters and actuator failures. IEEE Trans Fuzzy Syst 20(3):471–486
80. Gao Z, Ding SX (2007) State and disturbance estimator for time-delay systems with application to fault estimation and signal compensation. IEEE Trans Signal Process 55(12):5541–5551
81. Gao Z, Shi X, Ding SX (2008) Fuzzy state-disturbance observer design for T-S fuzzy systems with application to sensor fault estimation. IEEE Trans Syst Man Cybern, Part B-Cybern 38(3):875–880
82. Liu M, Cao X, Shi P (2013) Fault estimation and tolerant control for fuzzy stochastic systems. IEEE Trans Fuzzy Syst 21(1):221–229
83. Li HY, Chen ZR, Wu LG, Lam HK, Du HP (2017) Event-triggered fault detection of nonlinear networked systems. IEEE Trans Cybern 47(4):1041–1052
84. Ning ZK, Yu J, Pan YN, Li HY (2018) Adaptive event-triggered fault detection for fuzzy stochastic systems with missing measurements. IEEE Trans Fuzzy Syst 26(4):2201–2212
85. Liu Y, Yang GH, Li XJ (2017) Event-triggered fault detection observer design for affine fuzzy systems. Neurocomputing 267(6):564–571
86. Wang XL, Yang GH (2018) Event-triggered fault detection for discrete-time T-S fuzzy systems. ISA Trans 75:18–30
87. Peng C, Yang MJ, Zhang J, Fei MR, Hu SL (2017) Network-based H_∞ control for T-S fuzzy systems with an adaptive event-triggered communication scheme. Fuzzy Sets Syst 329:61–76
88. Liu YJ, Guo BZ, Park JH, Lee S (2018) Event-based reliable dissipative filtering for T-S fuzzy systems with asynchronous constraints. IEEE Trans Fuzzy Syst 26(4):2089–2098
89. Oliveira M, Geromel JC, Bernussou J (2002) Extended H 2 and H norm characterizations and controller parametrizations for discrete-time systems. Int J Control 75(9):666–679

Chapter 2
Piecewise Fuzzy H_∞ Filtering for T-S Fuzzy Systems

2.1 Introduction

The filtering of dynamic systems has become a hot topic in current research, and a large number of filter design methods have emerged. It should be noted that these research results are obtained based on the Lyapunov function analysis method, which is considered a powerful tool to ensure the stability of the systems [1, 2]. Moreover, the filtering problems for T-S fuzzy systems has also been widely studied within the framework of Lyapunov theory. For example, Refs. [3, 4]. However, it should be explained that the results mentioned above were presented based on the quadratic Lyapunov functions, which have been proved to be some conservatism. Therefore, it is necessary to explore new Lyapunov functions to reduce the conservatism.

Recently, some new types of Lyapunov functions have been established for the T-S fuzzy systems to relax the conservatism, including fuzzy Lyapunov functions [5–8], polynomial Lyapunov functions [9, 10] and piecewise Lyapunov functions [11–13], etc. Moreover, a special class of Lyapunov functions without monotonically decreasing were proposed in Refs. [14, 15], which have proved to be an effective strategy to reduce the conservatism. In Ref. [16], the state feedback H_∞ control strategy was explored by establishing a nonmonotonic Lyapunov function. However, it should be pointed out that the system states are available in [16], and the premise variables of the non-PDC controller are chosen as the same as those of the T-S fuzzy systems. It is well known that the internal states in many practical systems are not measurable, which leads to the unavailability of state information, and in this case, the method proposed in Ref. [16] will be invalidation. By establishing a Lyapunov function consisting of any number of quadratic functions, the stability problem of linear systems was investigated in Ref. [17], in which the higher-order derivatives of system state vectors in the continuous-time cases and the higher-order differences of system state vectors in discrete-time cases were applied respectively. However, it is noted that the stability problem of linear systems was studied in Ref. [17], while the H_∞ control problem cannot be solved within the theoretical framework of this

X.-L. Wang et al., *Robust Filtering and Fault Detection for T-S Fuzzy Systems*,
https://doi.org/10.1007/978-981-96-5818-3_2

reference. To sum up, in the aforementioned Refs. [16, 17], the approach presented in Ref. [16] cannot be used to deal with the situation where the system states are unknown, and the proposed method in Ref. [17] will fail in the presence of external disturbances in the systems. Therefore, it is very important to study the H_∞ control strategy for the T-S fuzzy systems with immeasurable state variables by proposing a new structured Lyapunov function.

This chapter is concerned with the design of H_∞ filers for T-S fuzzy systems. The conservatism of the existing results is reduced by establishing a new structured Lyapunov function, and the proposed method can solve the H_∞ control problem that cannot be dealt with in the existing result in Ref. [17]. By decomposing the space of premise variables into crisp regions and fuzzy regions, piecewise fuzzy H_∞ filters are established at different instants to cope with the challenges caused by the mismatch of the premise variables and exogenous disturbances. Based on the proposed new Lyapunov functions that depend on the different regions where the premise variables are located, sufficient conditions in terms of linear matrix inequalities are obtained to guarantee the stability and desired performance of the filtering error systems. Compared with the existing design methods, the technical solution proposed in this chapter can achieve better results.

2.2 System Modeling and Problem Statement

A class of discrete-time T-S fuzzy systems is considered, as shown below:

$$
\begin{aligned}
x(k+1) &= \sum_{i=1}^{r} h_i(g(x(k)))(A_i x(k) + D_i w(k)) \\
y(k) &= Cx(k) \\
z(k) &= \sum_{i=1}^{r} h_i(g(x(k)))E_i x(k)
\end{aligned}
\tag{2.1}
$$

in which $g(x(k)) = \begin{bmatrix} g_1(x(k)) \; g_2(x(k)) \; \cdots \; g_p(x(k)) \end{bmatrix}$ and $i \in \mathscr{E} \triangleq \{1, 2, \ldots, r\}$ stands for the premise variable vector and the ith fuzzy inference rule, respectively. $x(k) \in R^n$ is the unmeasurable state vector, $w_k \in R^w$ is the disturbance signal belonging to $L_2[0, \infty)$, $y_k \in R^m$ and $z_k \in R^z$ are, respectively, the measurement output and the output to be estimated. A_i, D_i, C and E_i are known constant matrices.

In addition, $h_i(g(x(k)))$ represents the membership function, described as follows:

$$
\sum_{i=1}^{r} h_i(g(x(k))) = 1, h_i(g(x(k))) \geq 0, \forall i = 1, 2, \ldots, r. \tag{2.2}
$$

The internal states in many practical systems are not measurable, and only the outputs of the systems can be directly obtained, which will result in inconsistency between the premise variables in system model and the ones in piecewise fuzzy filter. Therefore, in order to facilitate the design of piecewise fuzzy filters, the partition approach is applied to deal with the problems caused by the mismatch of premise variables in this chapter.

First, the premise variable space $\Upsilon \in \mathcal{R}^s$ is decomposed into ϱ polyhedral regions according to the following criterion:

$$\partial\Upsilon_{\varsigma}^{\iota} = \left\{ g(x(k))|h_{\varsigma}(x(k)) = 1, 0 \leq h_{\varsigma}(x(k+\kappa)) \underset{0<|\kappa|<<1}{<} 1, \varsigma \in \mathscr{E} \right\} \tag{2.3}$$

in which ι is the set of the face indices of the polyhedral hull and satisfies $\partial\Upsilon_{\varsigma} = \cup_{\iota}(\partial\Upsilon_{\varsigma}^{\iota})$, and the polyhedral region Υ_l, $l \in \ell = \{1, 2, \ldots, \varrho\}$ (ℓ stands for the polyhedral region index set) satisfies the following relations:

$$\Upsilon_l \cap \Upsilon_s = \partial\Upsilon_{\varsigma}^{\iota}, l \neq s,\ l, s \in \ell,\ \varsigma \in \mathscr{E} \tag{2.4}$$

Since there is always $\sum_{m\in\Upsilon_l} h_m = 1$ in each region Υ_l, the T-S fuzzy system (2.1) is reconstructed in the following forms:

$$\begin{aligned} x(k+1) &= A_l(h_k)x(k) + D_l(h_k)w(k) \\ y(k) &= Cx(k) \\ z(k) &= E_l(h_k)x(k) \end{aligned} \tag{2.5}$$

where

$$\begin{aligned} A_l(h_k) &= \sum_{m\in\ell(l)} h_m(g(x(k)))A_m,\ D_l(h_k) = \sum_{m\in\ell(l)} h_m(g(x(k)))D_m, \\ E_l(h_k) &= \sum_{m\in\ell(l)} h_m(g(x(k)))E_m \end{aligned}$$

Moreover, $\mathscr{I}_1$ is described as all possible region transitions, as follows:

$$\mathscr{I}_1 = \{(l, i_1, \ldots, i_N)|g(x(k)) \in \Upsilon_l, g(x(k+1)) \in \Upsilon_{i_1}, \ldots, g(x(k+N)) \in \Upsilon_{i_N}\} \tag{2.6}$$

in which $l = i_t$, $t = 1, 2, \ldots, N$, means that the premise variables of the T-S fuzzy system are located in the same region at different instants. Otherwise, the premise variables of the T-S fuzzy syste will be converted to different regions at different instants.

According to the region transitions mentioned above, the T-S fuzzy system at different instants are constructed as follows:

$$\begin{aligned} x(k+t+1) &= A_{i_t}(h_{k+t})x(k+t) + D_{i_t}(h_{k+t})w(k+t) \\ y(k+t) &= Cx(k+t) \\ z(k+t) &= E_{i_t}(h_{k+t})x(k+t), \quad t = 1, \ldots, N \end{aligned} \tag{2.7}$$

where

$$A_{i_t}(h_{k+t}) = \sum_{m\in\ell(i_t)} h_m(g(x(k+t)))A_m,\ D_{i_t}(h_{k+t}) = \sum_{m\in\ell(i_t)} h_m(g(x(k+t)))D_m,$$
$$E_{i_t}(h_{k+N}) = \sum_{m\in\ell(i_t)} h_m(g(x(k+t)))E_m$$

Since the states of the piecewise fuzzy system are not measurable, and therefore the functions of filter states are chosen as the premise variables for piecewise fuzzy filters. Based on this view, the piecewise fuzzy filter is established, as shown below.

Filter Rule $\mathscr{E}^n$: If $g_1(\hat{x}(k))$ is ϵ_{n1}, $g_2(\hat{x}(k))$ is ϵ_{n2}, ..., $g_p(\hat{x}(k))$ is ϵ_{np}, then

$$\begin{aligned} \hat{x}(k+1) &= A_n\hat{x}(k) + L_s(y(k) - \hat{y}(k)) \\ \hat{y}(k) &= C\hat{x}(k) \\ \hat{z}(k) &= E_n\hat{x}(k) \end{aligned} \tag{2.8}$$

in which $g(\hat{x}_k) = \left[\, g_1(\hat{x}(k))\ g_2(\hat{x}(k)) \cdots g_p(\hat{x}(k)) \,\right]$ stands for the premise variable vector that depends on the states of filter. $\hat{y}(k) \in R^m$ and $\hat{z}(k) \in R^z$ are the corresponding estimations of the outputs, respectively. L_s is the gain to be determined.

Then, the piecewise fuzzy filter is established based on the introduction of partition method, as follows:

$$\begin{aligned} \hat{x}(k+1) &= A_s(\hat{h}_k)\hat{x}(k) + L_s(y(k) - \hat{y}(k)) \\ \hat{y}(k) &= C\hat{x}(k) \\ \hat{z}(k) &= E_s(\hat{h}_k)\hat{x}(k) \end{aligned} \tag{2.9}$$

where

$$A_s(\hat{h}_k) = \sum_{n\in\ell(s)} h_n(\hat{x}(k))A_n,\ E_s(\hat{h}_k) = \sum_{n\in\ell(s)} h_n(\hat{x}(k))E_n$$

Also define $\mathscr{I}_2$ as all possible region transitions, as shown below:

$$\mathscr{I}_2 = \{(s, j_1, \ldots, j_N) | g(\hat{x}(k)) \in \Upsilon_s, g(\hat{x}(k+1)) \in \Upsilon_{j_1}, \ldots, g(\hat{x}(k+N)) \in \Upsilon_{j_N}\} \tag{2.10}$$

where $s, j_1, \ldots, j_N \in \ell$.

In addition, piecewise fuzzy filters at different instants are designed based on the established regional transformation relationship, as shown below:

$$\begin{aligned}\hat{x}(k+t+1) &= A_{j_t}(\hat{h}_{k+t})\hat{x}(k+t) + L_{j_t}(y(k+t) - \hat{y}(k+t))\\ \hat{y}(k+t) &= C\hat{x}(k+t)\\ \hat{z}(k+t) &= E_{j_t}(\hat{h}_{k+t})\hat{x}(k+t), \quad t = 1, \ldots, N\end{aligned} \tag{2.11}$$

where

$$A_{j_t}(\hat{h}_{k+t}) = \sum_{n\in\ell(j_t)} h_n(\hat{x}(k+t))A_n,\ E_{j_t}(\hat{h}_{k+t}) = \sum_{n\in\ell(j_t)} h_n(\hat{x}(k+t))E_n$$

Remark 2.1 Note that the premise variables of the T-S fuzzy system are different from those of the piecewise fuzzy filters because the states of the system are not measurable, and the partition approach is applied to deal with the mismatch of the premise variables. In addition, in order to cope with the challenges caused by the exogenous disturbances in the case of the proposed special structured Lyapunov function, the piecewise fuzzy system and the piecewise fuzzy filter are, respectively, constructed as multiple system models at different instants.

Remark 2.2 The asynchronous problems of premise variables have been reported in Refs. [10, 18–20], in which it should be pointed out that the number of fuzzy rules in the filters established in Refs. [18, 19] is the same as that in the system models, and the number of fuzzy rules in the controller established in Refs. [10, 20] is different from that in the system models. In this chapter, the same number of fuzzy rules are considered.

Define $\xi(k) = \left[x^T(k)\ \tilde{x}^T(k)\right]^T$, $\xi(k+t) = \left[x^T(k+t)\ \tilde{x}^T(k+t)\right]^T$, $\tilde{x}(k) = x(k) - \hat{x}(k)$, $\tilde{x}(k+t) = x(k+t) - \hat{x}(k+t)$, $\tilde{z}(k) = z(k) - \hat{z}(k)$ and $\tilde{z}(k+t) = z(k+t) - \hat{z}(k+t)$, then the filtering error systems consisting of (2.5) and (2.9), (2.7) and (2.11) are constructed and presented as follows:

$$\begin{cases}\xi(k+1) = A_{ls}\xi(k) + D_{ls}w(k)\\ \tilde{z}(k) = E_{ls}\xi(k)\end{cases} \tag{2.12}$$

$$\begin{cases}\xi(k+t+1) = A_{i_tj_t}\xi(k+t) + D_{i_tj_t}w(k+t)\\ \tilde{z}(k+t) = E_{i_tj_t}\xi(k+t), \quad t = 1, \ldots, N\end{cases} \tag{2.13}$$

where

$$A_{ls} = \begin{bmatrix} A_l(h_k) & 0\\ A_l(h_k) - A_s(\hat{h}_k) & A_s(\hat{h}_k) - L_sC\end{bmatrix}, D_{ls} = \begin{bmatrix} D_l(h_k)\\ D_l(h_k)\end{bmatrix},$$

$$A_{i_tj_t} = \begin{bmatrix} A_{i_t}(h_{k+t}) & 0\\ A_{i_t}(h_{k+t}) - A_{j_t}(\hat{h}_{k+t}) & A_{j_t}(\hat{h}_{k+t}) - L_{j_t}C\end{bmatrix}, D_{i_tj_t} = \begin{bmatrix} D_{i_t}(h_{k+t})\\ D_{i_t}(h_{k+t})\end{bmatrix},$$

$$E_{ls} = \left[\, E_l(h_k) - E_s(\hat{h}_k)\ E_s(\hat{h}_k) \,\right], E_{i_t j_t} = \left[\, E_{i_t}(h_{k+t}) - E_{j_t}(\hat{h}_{k+t})\ E_{j_t}(\hat{h}_{k+t}) \,\right]$$

Problem: For the discrete-time T-S fuzzy system (2.1), and the piecewise T-S fuzzy systems (2.5) and (2.7), the purpose of this chapter is to construct the piecewise fuzzy filters (2.9) and (2.11), such that the following two performances are satisfied:

(S.1) *Asymptotic Stability:* The filtering error systems (2.12) and (2.13) with $w_k = 0$ and $w_{k+t} = 0$, $t = 1, \ldots, N$ are asymptotically stable.

(S.2) H_∞ *Performance:* The filtering error systems (2.12) and (2.13) satisfy the following inequality:

$$\sum_{t=1}^{N}(\|\tilde{z}_{k+t-1}\|_2) \leq \sum_{t=1}^{N}(\alpha\|w_{k+t-1}\|_2) \tag{2.14}$$

for all $w_k \in L_2[0, \infty)$ and $w_{k+t} \in L_2[0, \infty)$, $t = 1, \ldots, N$, in which α is the H_∞ performance index.

At the same time, four purposes need to be achieved. (1) The mismatch of premise variables incurred by the immeasurable state variables is solved in the piecewise fuzzy filters synthesis. (2) The disturbance signals w_k and w_{k+t} can be effectively suppressed, and the output signals $\tilde{z}_k$ and $\tilde{z}_{k+t}$ can be accurately estimated. (3) The Lyapunov function is handled respectively in different regions to avoid the emergence of multiple power of the augmented matrix. (4) The new design technology proposed in this chapter achieves better H_∞ performance than the existing design methods.

2.3 Piecewise Fuzzy H_∞ Filter Design

In this section, a new Lyapunov function is first put forward based on the partition regions of premise variables mentioned above, in which the Lyapunov function is made up of multiple sub-Lyapunov functions, and each sub-Lyapunov function is not required to be positive definite. Instead of that, it is only necessary that the new Lyapunov function is positive definite and its differential is negative definite. A lemma is given before presenting the main results, as follows.

Lemma 2.1 ([14]) *If there are continuous functions* $V_1(\xi_k)$, …, $V_N(\xi_k)$, *then the equation* $\sum_{t=1}^{N} tV_t(\xi_k) = 0$ *holds.* $\sum_{t=q}^{N} V_t(\xi_k) > 0, \forall \xi_k \neq 0,$ *and radially unbounded for* $q = 1, 2, \ldots, N$, *and*

$$\Delta_k V(\xi(k)) = \sum_{t=1}^{N}(V_t(\xi(k+t)) - V_t(\xi(k))) < 0$$

then, the origin is a globally asymptotically stable equilibrium of the system.

2.3.1 H_∞ Performance Analysis Conditions

Next, the performance analysis results of piecewise fuzzy filters for T-S fuzzy systems will be given.

Theorem 2.1 *For the T-S fuzzy system (2.1), and the piecewise T-S fuzzy systems (2.5) and (2.7), design piecewise fuzzy filters (2.9) and (2.11) respectively. The filtering error systems (2.12) and (2.13) satisfy the performance conditions (S.1) and (S.2) if there exist matrices $P_{i_t j_t}$, P_{ls}, $t = 1, 2, \ldots, N$, $\alpha > 0$, $Q_{i_1 j_1}$, $Q_{i_2 j_2}$, ..., $Q_{i_{N-1} j_{N-1}}$ such that the following inequalities hold with $l, s \in \ell$, $(l, i_1, i_2, \ldots, i_N) \in \mathscr{I}_1$, $(s, j_1, j_2, \ldots, j_N) \in \mathscr{I}_2$.*

$$\sum_{t=q}^{N} P_{ls}^{t} > 0, q = 1, 2, \ldots, N \tag{2.15}$$

$$\begin{bmatrix} 1^N & 0 & A_{i_{N-1} j_{N-1}}^T P_{i_N j_N}^N & E_{i_{N-1} j_{N-1}}^T \\ \star & -\alpha^2 I & D_{i_{N-1} j_{N-1}}^T P_{i_N j_N}^N & 0 \\ \star & \star & -P_{i_N j_N}^N & 0 \\ \star & \star & \star & -I \end{bmatrix} < 0 \tag{2.16}$$

$$\begin{bmatrix} 1^{N-1} & 0 & A_{i_{N-2} j_{N-2}}^T Q_{i_{N-1} j_{N-1}}^{N-1} & 2^{N-1} \\ \star & -\alpha^2 I & D_{i_{N-2} j_{N-2}}^T Q_{i_{N-1} j_{N-1}}^{N-1} & 0 \\ \star & \star & -Q_{i_{N-1} j_{N-1}}^{N-1} & 0 \\ \star & \star & \star & -I \end{bmatrix} < 0 \tag{2.17}$$

$$\vdots \tag{2.18}$$

$$\begin{bmatrix} -\sum_{t=1}^{N} P_{ls}^{t} & 0 & A_{ls}^T Q_{i_1 j_1}^1 & E_{ls}^T \\ \star & -\alpha^2 I & D_{ls}^T Q_{i_1 j_1}^1 & 0 \\ \star & \star & -Q_{i_1 j_1}^1 & 0 \\ \star & \star & \star & -I \end{bmatrix} < 0 \tag{2.19}$$

in which

$$1^N = P_{i_{N-1} j_{N-1}}^{N-1} - Q_{i_{N-1} j_{N-1}}^{N-1}, \; 1^{N-1} = P_{i_{N-2} j_{N-2}}^{N-2} - Q_{i_{N-2} j_{N-2}}^{N-2}, \; 2^{N-1} = E_{i_{N-2} j_{N-2}}^T$$

Proof Different from the existing Lyapunov function in Ref. [17], a new piecewise fuzzy Lyapunov function consisting of multiple piecewise quadratic Lyapunov functions is constructed, in which the augmented vectors in each piecewise quadratic Lyapunov function are distributed in different partition regions. Next, the Lyapunov function is presented as follows:

$$V^k(\xi(k)) = \sum_{t=1}^{N} \sum_{q=t}^{N} V_q^{k+t-1} \tag{2.20}$$

and

$$
\begin{aligned}
V_q^k(\xi(k)) &= \xi(k)^T P_{ls}^q \xi(k), \\
V_q^{k+t}(\xi(k+t)) &= \xi(k+t)^T P_{i_t j_t}^q \xi(k+t)
\end{aligned}
\tag{2.21}
$$

in which $P_{ls}^q, P_{i_t j_t}^q, l, s, i_t, j_t \in \ell = \{1, 2, \ldots, \varrho\}, t, q = 1, 2, \ldots, N$, are matrices to be determined.

From (2.20), it is known that if the inequality (2.15) is established, and then the new Lyapunov function (2.20) will satisfy $\sum_{t=1}^{N} V_t(\xi_k) > 0$.

Next, the differential equation is further presented on the basis of (2.20) and (2.21), as shown below:

$$
\begin{aligned}
\Delta_k V(\xi(k)) &= \sum_{t=1}^{N} (\xi(k+t)^T P_{i_t j_t}^t \xi(k+t) - \xi(k)^T P_{ls}^t \xi(k)) \\
&= (V_N^{k+N} - V_N^k) + (V_{N-1}^{k+N-1} - V_{N-1}^k) + \cdots + (V_1^{k+1} - V_1^k)
\end{aligned}
\tag{2.22}
$$

Then, according to the system models (2.12) and (2.13), one has

$$V_1^{k+1} - V_1^k = \xi(k+1)^T P_{i_1 j_1}^1 \xi(k+1) - \xi(k)^T P_{ls}^1 \xi(k) \tag{2.23}$$

$$V_2^{k+2} - V_2^k = \xi(k+2)^T P_{i_2 j_2}^2 \xi(k+2) - \xi(k)^T P_{ls}^2 \xi(k) \tag{2.24}$$

$$\vdots \tag{2.25}$$

$$V_N^{k+N} - V_N^k = \xi(k+N)^T P_{i_N j_N}^N \xi(k+N) - \xi(k)^T P_{ls}^N \xi(k) \tag{2.26}$$

In addition, in order to simplify the notation, the following equation is defined:

$$
\begin{aligned}
J = \sum_{t=1}^{N} (&\xi(k+t)^T P_{i_t j_t}^t \xi(k+t) - \xi(k)^T P_{ls}^t \xi(k) \\
&+ \tilde{z}(k+t-1)^T \tilde{z}(k+t-1) - \alpha^2 w(k+t-1)^T w(k+t-1))
\end{aligned}
\tag{2.27}
$$

Next, the Lyapunov function is handled respectively in different regions to avoid generating the multiple power of the augmented matrix, and the partition proof steps will be presented, respectively, as shown below:

Step 1: Adding and subtracting $\xi_{k+N-1}^T Q_{i_{N-1} j_{N-1}}^{N-1} \xi_{k+N-1}$ to (2.27), one has

$$
\begin{aligned}
J = &\ \xi(k+N)^T P_{i_N j_N}^N \xi(k+N) + \xi(k+N-1)^T P_{i_{N-1} j_{N-1}}^{N-1} \xi(k+N-1) \\
&- \xi(k+N-1)^T Q_{i_{N-1} j_{N-1}}^{N-1} \xi(k+N-1) + \tilde{z}(k+N-1)^T \tilde{z}(k+N-1) \\
&- \alpha^2 w(k+N-1)^T w(k+N-1) + \xi(k+N-1)^T Q_{i_{N-1} j_{N-1}}^{N-1} \xi(k+N-1) \\
&+ \sum_{t=1}^{N-2} (\xi(k+t)^T P_{i_t j_t}^t \xi(k+t)) + \sum_{t=1}^{N-1} (\tilde{z}(k+t-1)^T \tilde{z}(k+t-1)
\end{aligned}
\tag{2.28}
$$

$$- \alpha^2 w(k+t-1)^T w(k+t-1)) - \sum_{t=1}^{N} (\xi(k)^T P_{ls}^t \xi(k))$$

According to the system model (2.13) with $t = N$, one has

$$\begin{cases} \xi(k+N) = A_{i_{N-1} j_{N-1}} \xi(k+N-1) + D_{i_{N-1} j_{N-1}} w(k+N-1) \\ \tilde{z}(k+N-1) = E_{i_{N-1} j_{N-1}} \xi(k+N-1) \end{cases}$$

and then

$$\begin{aligned} J = {} & (A_{i_{N-1} j_{N-1}} \xi(k+N-1) + D_{i_{N-1} j_{N-1}} w(k+N-1))^T P_{i_N j_N}^N \\ & \times (A_{i_{N-1} j_{N-1}} \xi(k+N-1) + D_{i_{N-1} j_{N-1}} w(k+N-1)) \\ & + \xi(k+N-1)^T P_{i_{N-1} j_{N-1}}^{N-1} \xi(k+N-1) \\ & - \xi(k+N-1)^T Q_{i_{N-1} j_{N-1}}^{N-1} \xi(k+N-1) \\ & + (E_{i_{N-1} j_{N-1}} \xi(k+N-1))^T (E_{i_{N-1} j_{N-1}} \xi(k+N-1)) \\ & - \alpha^2 w(k+N-1)^T w(k+N-1) \\ & + \xi(k+N-1)^T Q_{i_{N-1} j_{N-1}}^{N-1} \xi(k+N-1) + \sum_{t=1}^{N-2} (\xi(k+t)^T P_{i_t j_t}^t \xi(k+t)) \\ & + \sum_{t=1}^{N-1} (\tilde{z}(k+t-1)^T \tilde{z}(k+t-1) - \alpha^2 w(k+t-1)^T w(k+t-1)) \\ & - \sum_{t=1}^{N} (\xi(k)^T P_{ls}^t \xi(k)) \end{aligned} \tag{2.29}$$

Using the inequality (2.16) in Theorem 2.1, which is pre- and post-multiplied by $[\xi_{k+N-1}^T, w_{k+N-1}^T]$ and its transpose, to get

$$\begin{aligned} & (A_{i_{N-1} j_{N-1}} \xi(k+N-1) + D_{i_{N-1} j_{N-1}} w(k+N-1))^T P_{i_N j_N}^N \\ & \times (A_{i_{N-1} j_{N-1}} \xi(k+N-1) + D_{i_{N-1} j_{N-1}} w(k+N-1)) \\ & + \xi(k+N-1)^T P_{i_{N-1} j_{N-1}}^{N-1} \xi(k+N-1) \\ & - \xi(k+N-1)^T Q_{i_{N-1} j_{N-1}}^{N-1} \xi(k+N-1) \\ & + (E_{i_{N-1} j_{N-1}} \xi(k+N-1))^T (E_{i_{N-1} j_{N-1}} \xi(k+N-1)) \\ & - \alpha^2 w(k+N-1)^T w(k+N-1) \\ & < 0 \end{aligned} \tag{2.30}$$

Substituting (2.30) into (2.29) yields:

$$J < \xi(k+N-1)^T Q^{N-1}_{i_{N-1}j_{N-1}} \xi(k+N-1) + \sum_{t=1}^{N-2} (\xi(k+t)^T P^t_{i_t j_t} \xi(k+t)) \quad (2.31)$$
$$+ \sum_{t=1}^{N-1} (\tilde{z}(k+t-1)^T \tilde{z}(k+t-1) - \alpha^2 w(k+t-1)^T w(k+t-1))$$
$$- \sum_{t=1}^{N} (\xi(k)^T P^t_{ls} \xi(k))$$

Step 2: Adding and subtracting $\xi^T_{k+N-2} Q^{N-2}_{i_{N-2}j_{N-2}} \xi_{k+N-2}$ to (2.31), one has

$$\begin{aligned} J <\ & \xi(k+N-1)^T Q^{N-1}_{i_{N-1}j_{N-1}} \xi(k+N-1) + \xi(k+N-2)^T P^{N-2}_{i_{N-2}j_{N-2}} \xi(k+N-2) \\ & + \tilde{z}(k+N-2)^T \tilde{z}(k+N-2) - \alpha^2 w(k+N-2)^T w(k+N-2) \\ & - \xi(k+N-2)^T Q^{N-2}_{i_{N-2}j_{N-2}} \xi(k+N-2) + \xi(k+N-2)^T Q^{N-2}_{i_{N-2}j_{N-2}} \xi(k+N-2) \\ & + \sum_{t=1}^{N-3} (\xi(k+t)^T P^t_{i_t j_t} \xi(k+t)) + \sum_{t=1}^{N-2} (\tilde{z}(k+t-1)^T \tilde{z}(k+t-1) \\ & - \alpha^2 w(k+t-1)^T w(k+t-1)) - \sum_{t=1}^{N} (\xi(k)^T P^t_{ls} \xi(k)) \end{aligned} \quad (2.32)$$

According to the system model (2.13) with $t = N-1$, one has

$$\begin{cases} \xi(k+N-1) = A_{i_{N-2}j_{N-2}} \xi(k+N-2) + D_{i_{N-2}j_{N-2}} w(k+N-2) \\ \tilde{z}(k+N-2) = E_{i_{N-2}j_{N-2}} \xi(k+N-2) \end{cases}$$

Then, the inequality (2.32) is rewritten as follows:

$$\begin{aligned} J <\ & (A_{i_{N-2}j_{N-2}} \xi(k+N-2) + D_{i_{N-2}j_{N-2}} w(k+N-2))^T Q^{N-1}_{i_{N-1}j_{N-1}} \\ & \times (A_{i_{N-2}j_{N-2}} \xi(k+N-2) + D_{i_{N-2}j_{N-2}} w(k+N-2)) \\ & + \xi(k+N-2)^T P^{N-2}_{i_{N-2}j_{N-2}} \xi(k+N-2) \\ & + (E_{i_{N-2}j_{N-2}} \xi(k+N-2))^T (E_{i_{N-2}j_{N-2}} \xi(k+N-2)) \\ & - \alpha^2 w(k+N-2)^T w(k+N-2) - \xi(k+N-2)^T Q^{N-2}_{i_{N-2}j_{N-2}} \xi(k+N-2) \\ & + \xi(k+N-2)^T Q^{N-2}_{i_{N-2}j_{N-2}} \xi(k+N-2) + \sum_{t=1}^{N-3} (\xi(k+t)^T P^t_{i_t j_t} \xi(k+t)) \\ & + \sum_{t=1}^{N-2} (\tilde{z}(k+t-1)^T \tilde{z}(k+t-1) - \alpha^2 w(k+t-1)^T w(k+t-1)) \\ & - \sum_{t=1}^{N} (\xi(k)^T P^t_{ls} \xi(k)) \end{aligned} \quad (2.33)$$

Using the inequality (2.17) in Theorem 2.1, which is pre- and post-multiplied by $[\xi_{k+N-2}^T, w_{k+N-2}^T]$ and its transpose, to get

$$\begin{aligned}
&(A_{i_{N-2}j_{N-2}}\xi(k+N-2) + D_{i_{N-2}j_{N-2}}w(k+N-2))^T Q_{i_{N-1}j_{N-1}}^{N-1} \\
&\times (A_{i_{N-2}j_{N-2}}\xi(k+N-2) + D_{i_{N-2}j_{N-2}}w(k+N-2)) \\
&+ \xi(k+N-2)^T P_{i_{N-2}j_{N-2}}^{N-2}\xi(k+N-2) \\
&+ (E_{i_{N-2}j_{N-2}}\xi(k+N-2))^T (E_{i_{N-2}j_{N-2}}\xi(k+N-2)) \\
&- \alpha^2 w(k+N-2)^T w(k+N-2) \\
&- \xi(k+N-2)^T Q_{i_{N-2}j_{N-2}}^{N-2}\xi(k+N-2) \\
&< 0
\end{aligned} \tag{2.34}$$

Then, substituting (2.34) into (2.33) yields:

$$\begin{aligned}
J <&\xi(k+N-2)^T Q_{i_{N-2}j_{N-2}}^{N-2}\xi(k+N-2) + \sum_{t=1}^{N-3}(\xi(k+t)^T P_{i_t j_t}^t \xi(k+t)) \\
&+ \sum_{t=1}^{N-2}(\tilde{z}(k+t-1)^T\tilde{z}(k+t-1) - \alpha^2 w(k+t-1)^T w(k+t-1)) \\
&- \sum_{t=1}^{N}(\xi(k)^T P_{l_s}^t \xi(k))
\end{aligned} \tag{2.35}$$

Repeat the above steps $k-3$ times, and then one has

$$J < \xi(k+1)^T Q_{i_1 j_1}^1 \xi(k+1) + \tilde{z}(k)^T\tilde{z}(k) - \alpha^2 w(k)^T w(k) - \sum_{t=1}^{N}(\xi(k)^T P_{l_s}^t \xi(k)) \tag{2.36}$$

substituting (2.12) into (2.36), and using the inequality (2.19) in Theorem 2.1, which is pre- and post-multiplied by $[\xi_k^T, w_k^T]$ and its transpose, to get

$$\xi(k+1)^T Q_{i_1 j_1}^1 \xi(k+1) + \tilde{z}(k)^T\tilde{z}(k) - \alpha^2 w(k)^T w(k) - \sum_{t=1}^{N}(\xi(k)^T P_{l_s}^t \xi(k)) < 0 \tag{2.37}$$

After the above analysis, it is indicated from (2.28) to (2.37) that the inequality $J = \sum_{t=1}^{N}(\xi_{k+t}^T P_{i_t j_t}^t \xi_{k+t} - \xi_k^T P_{l_s}^t \xi_k + \tilde{z}_{k+t-1}^T \tilde{z}_{k+t-1} - \alpha^2 w_{k+t-1}^T w_{k+t-1}) < 0$ holds. Furthermore, by Lemma 2.1, it is concluded that the inequality conditions (2.15)–(2.19) can ensure the stability and H_∞ performance of the filtering error systems. The proof is completed. □

Remark 2.3 Unlike the Lyapunov function constructed in Ref. [17], a new Lyapunov function which depends on the regions of premise variables and consists of multiple sub-Lyapunov functions is proposed in this chapter, in which all the sub-Lyapunpv functions $V_q^{k+t-1}(\xi_{k+t-1})$ are quadratic in the states of the filtering error systems, and each sub-Lyapunpv function is not required to be positive.

Remark 2.4 It should be noted that the new Lyapunov function established based on the partition regions of the premise variables is difficult to be dealt with, and the introduction of the new Lyapunov function also result in the generation of multiple power of the augmented matrix, for example, $A_{ls}^T A_{i_1 j_1}^T \cdots A_{i_N j_N}^T P_{i_N j_N}^N A_{i_N j_N} \cdots A_{i_1 j_1} A_{ls}$ and $D_{ls}^T A_{i_1 j_1}^T \cdots A_{i_N j_N}^T P_{i_N j_N}^N A_{i_N j_N} \cdots A_{i_1 j_1} D_{ls}$, which will be an unsolvable problem. In order to overcome this difficulty, the Lyapunov function is handled respectively in different regions to avoid the emergence of multiple power of the augmented matrix.

Compared with the existing Lyapunov functions, the Lyapunov function constructed in this chapter is more general, and the existing Lyapunov functions are considered as special cases of the established Lyapunov function. For example, when $N = 1$, the designed method is reduced to the existing results in Refs. [11, 19, 21]. Moreover, the following corollary is constructed in the case of $N = 1$.

Corollary 2.1 *For the T-S fuzzy system (2.1), and the piecewise T-S fuzzy systems (2.5), design piecewise fuzzy filter (2.9). The filtering error system (2.12) satisfies the performance conditions (S.1) and (S.2) if there exist matrices* $P_{i_1 j_1} > 0$, $P_{ls} > 0$, X_s, $\alpha > 0$ *such that the following inequality holds with* $l, s \in \ell$, $(l, i_1) \in \mathscr{I}_1$, $(s, j_1) \in \mathscr{I}_2$.

$$\begin{bmatrix} -P_{ls} & 0 & A_{ls}^T P_{i_1 j_1} & E_{ls}^T \\ \star & -\alpha^2 I & D_{ls}^T P_{i_1 j_1} & 0 \\ \star & \star & -P_{i_1 j_1} & 0 \\ \star & \star & \star & -I \end{bmatrix} < 0$$

Proof The proof is similar to that in Refs. [11, 19, 21], so omitted here. □

2.3.2 Piecewise Fuzzy H_∞ Filter Design Conditions

It should be pointed out that in Theorem 2.1, the matrices $P_{i_t j_t}^N$, $Q_{i_{N-1} j_{N-1}}^{N-1}$, ..., $Q_{i_1 j_1}^1$ are coupled with the matrices $A_{i_{t-1} j_{t-1}}^T$, $D_{i_{t-1} j_{t-1}}^T$, $A_{i_{t-2} j_{t-2}}^T$, $D_{i_{t-2} j_{t-2}}^T$, A_{ls}^T and D_{ls}^T, $t = 1, 2, \ldots, N$, respectively. Thus, the inequality conditions (2.15)–(2.19) are nonconvex and cannot be solved directly. Therefore, the linear matrix inequality conditions satisfying the performances (S.1) and (S.2) of the filtering error systems will be presented, as shown below.

Theorem 2.2 *For the T-S fuzzy system (2.1), and the piecewise T-S fuzzy systems (2.5) and (2.7), piecewise fuzzy filters (2.9) and (2.11) are designed, respectively.*

The filtering error systems (2.12) and (2.13) satisfy the performance conditions (S.1) and (S.2) if there exist matrices $P_{i_t j_t}$, P_{ls}, X_s, X_{j_t}, $t = 1, 2, \ldots, N$, $Q_{i_1 j_1}$, $Q_{i_2 j_2}$, $\ldots$, $Q_{i_{N-1} j_{N-1}}$, $R = \begin{bmatrix} R_{11} & R_{12} \\ R_{22} & R_{22} \end{bmatrix}$, *in which* $R_{22} > 0$, $\alpha > 0$, *such that the following inequalities hold with* $l, s \in \ell$, $(l, i_1, i_2, \ldots, i_N) \in \mathscr{I}_1$, $(s, j_1, j_2, \ldots, j_N) \in \mathscr{I}_2$.

$$\sum_{t=q}^{N} P_{ls}^t > 0, q = 1, 2, \ldots, N \tag{2.38}$$

$$\begin{bmatrix} 1^N & 0 & A_{i_{N-1} j_{N-1}}^T R & E_{i_{N-1} j_{N-1}}^T \\ \star & -\alpha^2 I & D_{i_{N-1} j_{N-1}}^T R & 0 \\ \star & \star & P_{i_N j_N}^N - R - R^T & 0 \\ \star & \star & \star & -I \end{bmatrix} < 0 \tag{2.39}$$

$$\begin{bmatrix} 1^{N-1} & 0 & A_{i_{N-2} j_{N-2}}^T R & 2^{N-1} \\ \star & -\alpha^2 I & D_{i_{N-2} j_{N-2}}^T R & 0 \\ \star & \star & Q_{i_{N-1} j_{N-1}}^{N-1} - R - R^T & 0 \\ \star & \star & \star & -I \end{bmatrix} < 0 \tag{2.40}$$

$$\vdots \tag{2.41}$$

$$\begin{bmatrix} -\sum_{t=1}^N P_{ls}^t & 0 & A_{ls}^T R & E_{ls}^T \\ \star & -\alpha^2 I & D_{ls}^T R & 0 \\ \star & \star & Q_{i_1 j_1}^1 - R - R^T & 0 \\ \star & \star & \star & -I \end{bmatrix} < 0 \tag{2.42}$$

Then, the gains of the piecewise fuzzy filters are obtained by the following calculation:

$$L_s = (X_s R_{22}^{-1})^T, L_{j_t} = (X_{j_t} R_{22}^{-1})^T, t = 1, 2, \ldots, N \tag{2.43}$$

Proof A new matrix R is first introduced and presented as $R = \begin{bmatrix} R_{11} & R_{12} \\ R_{22} & R_{22} \end{bmatrix}$, in which $R_{22} > 0$. Then, by performing the congruence transformation $diag\{I, I, Y, I\}$ to the inequalities (2.16)–(2.19), and define $L_s^T Y$ and $L_{j_t}^T Y$ as new variables X_s and X_{j_t}, according to the inequality conditions: $-R^T P_{i_t j_t}^{-1} R \leq P_{i_t j_t} - R - R^T$, $t = 1, 2, \ldots, N$, $-R^T Q_{i_{N-1} j_{N-1}}^{-1} R \leq Q_{i_{N-1} j_{N-1}} - R - R^T$, $\ldots$, and $-R^T Q_{i_1 j_1}^{-1} R \leq Q_{i_1 j_1} - R - R^T$, the matrix inequality conditions (2.16)–(2.19) in Theorem 2.1 are linearized to the inequalities (2.39)–(2.42). Thus, the proof is completed. □

Remark 2.5 In Ref. [17], a new type of Lyapunov function was investigated. However, the stability problem of linear systems was studied in Ref. [17], and the method in Ref. [17] will no longer be applicable in the presence of external disturbances in the system. In this chapter, by establishing piecewise fuzzy H_∞ filters at different instants to cope with the challenges caused by the exogenous disturbances, and the new Lyapunov function is handled respectively in different regions to avoid the emergence of multiple power of the augmented matrix, the H_∞ control problem of the T-S fuzzy systems is solved.

Remark 2.6 Note that the more sub-Lyapunov functions are included in the proposed Lyapunov function, the less conservative conditions are obtained, and the computational complexity will be increased. Therefore, there should be a trade-off between the improvement system performances and the computational burdens.

2.4 Numeral Example

In this section, the new design technology proposed in this chapter is verified through the tunnel diode circuit in Ref. [22]. The tunnel diode circuit diagram is shown in Fig. 2.1, and the relevant parameters are presented as: $C = 20\,\text{mf}$ $L = 1000\,\text{mH}$ $R = 10\,\Omega$.

Next, $x_{1k} = v_{ck}$ and $x_{2k} = i_{Lk}$ are defined as the state variables respectively, and then the T-S fuzzy system is constructed by setting the sampling period as $T_s = 20\,\text{ms}$, as shown below:

$$\begin{aligned} x(k+1) &= \sum_{i=1}^{2} h_i(g(x(k)))(A_i x(k) + D_i w(k)) \\ y(k) &= Cx(k) \\ z(k) &= \sum_{i=1}^{2} h_i(g(x(k)))E_i x(k) \end{aligned}$$

in which

$$A_1 = \begin{bmatrix} 0.9887 & 0.9024 \\ -0.0180 & 0.8100 \end{bmatrix}, A_2 = \begin{bmatrix} 0.9034 & 0.8617 \\ -0.0172 & 0.8103 \end{bmatrix}, D_1 = \begin{bmatrix} 0.0093 \\ 0.0181 \end{bmatrix},$$

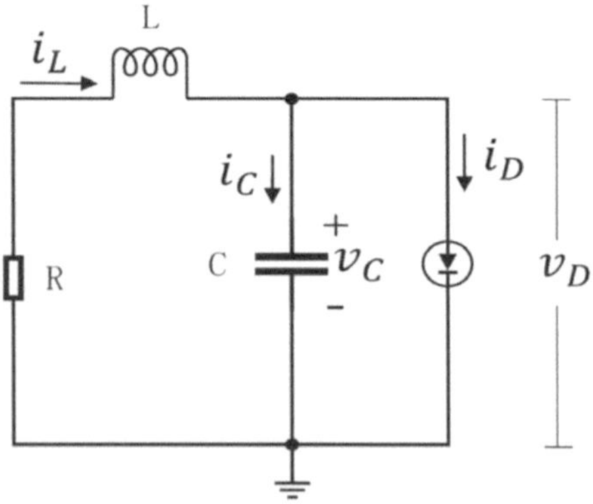

Fig. 2.1 Tunnel diode circuit

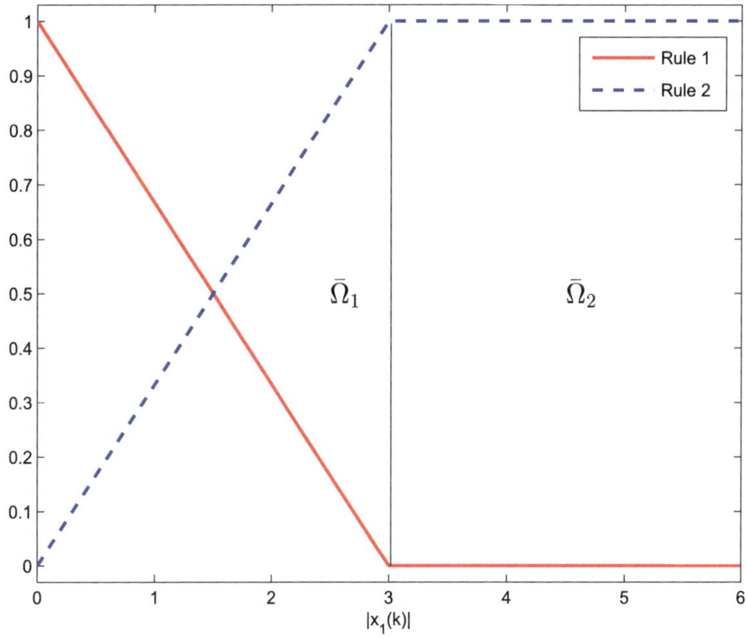

Fig. 2.2 Membership functions

$$D_2 = \begin{bmatrix} 0.0091 \\ 0.0181 \end{bmatrix}, C = \begin{bmatrix} 1 & 0 \end{bmatrix}, E_1 = \begin{bmatrix} 1 & 0 \end{bmatrix}, E_2 = \begin{bmatrix} 1 & 0 \end{bmatrix}$$

Based on the partition method introduced in this chapter, the premise variable space is decomposed into two regions, including $\bar{\Omega}_1$ and $\bar{\Omega}_2$, and presented in Fig. 2.2. In addition, the functions of the fuzzy filter states are selected as the premise variables of the piecewise fuzzy filter.

The purpose of this chapter is to find the minimum value of $\alpha > 0$, such that the inequality conditions established in Theorem 2.2 can guarantee the filtering error systems to achieve the asymptotic stability and the better H_∞ performance. Three cases are considered respectively to demonstrate the superiority of the designed method, as follows:

Case 1: $V^k(\xi_k) = \sum_{t=1}^{N} \sum_{q=t}^{N} V_q^{k+t-1}$ with $N = 1$.
Case 2: $V^k(\xi_k) = \sum_{t=1}^{N} \sum_{q=t}^{N} V_q^{k+t-1}$ with $N = 2$.
Case 3: $V^k(\xi_k) = \sum_{t=1}^{N} \sum_{q=t}^{N} V_q^{k+t-1}$ with $N = 3$.

in which Case 1 is the result of the degradation of the proposed method to the existing Refs. [11, 19, 21]. By solving the inequalities constructed in Theorem 2.2, the minimum α in three cases are obtained respectively and presented in Table 2.1. It is shown from Table 2.1 that the more sub-Lyapunov functions are included in the Lyapunov function, the smaller α are obtained.

Table 2.1 The minimum α

Index	Theorem 2.2 with N = 2	Theorem 2.2 with N = 3	The methods [11]
α	1.5861	0.7561	6.8942

In addition, Case 3 is selected in this example to further illustrate the effectiveness of the new design technology. First, define the parameter $N = 3$ and then use the Matlab Toolbox to solve the inequalities (2.38)–(2.42) with $i_3 = 1, 2, i_2 = 1, 2, i_1 = 1, 2, j_3 = 1, 2, j_2 = 1, 2, j_1 = 1, 2, l = 1, 2, s = 1, 2$. Furthermore, the filter gains can be calculated according to the Eq. (2.43) as follows:

$$L_1 = \begin{bmatrix} 1.0926 \\ 0.1560 \end{bmatrix}, L_2 = \begin{bmatrix} 1.0925 \\ 0.1561 \end{bmatrix}$$

Assume that the disturbance $w(k) = \frac{2}{1+0.8k}$ enters into the T-S fuzzy system at the beginning. Then, based on the obtained filter gains, the states of the system and the filter are shown in Figs. 2.3 and 2.4. The estimation errors of outputs $z(k)$ with $N = 1, 2, 3$ are shown in Fig. 2.5, and it is shown from Fig. 2.5 that the outputs of the T-S fuzzy system can be estimated accurately by the piecewise fuzzy filters. Furthermore, it is concluded that the more sub-Lyapunov functions are included in the new Lyapunov function, the smaller estimation errors are obtained, which further verifies the effectiveness of the proposed new design technology.

In addition, each piecewise quadratic Lyapunov function and the proposed new Lyapunov function are presented in Figs. 2.6, 2.7, 2.8 and 2.9, respectively. It is indicated from Figs. 2.6, 2.7, 2.8 and 2.9 that the sub-Lyapunov function V_2 is negative definite, the sub-Lyapunov function V_1 is not always positive definite, the sub-Lyapunov function V_3 and the Lyapunov functional V are positive definite. Therefore, it is concluded that the proposed new design technology in this paper is completely effective.

2.5 Summary

This chapter has investigated the H_∞ filter design problem for the T-S fuzzy system with immeasurable state variables. The mismatch of the premise variables caused by unknown system states is addressed by decomposing the space of premise variables into crisp regions and fuzzy regions. Then, according to these decomposed regions, piecewise fuzzy H_∞ filters are established at different instants to copy with the exogenous disturbances. Furthermore, by establishing new Lyapunov function, some sufficient conditions are obtained to ensure that the filtering error systems satisfy two performances. Finally, the design technology proposed in this chapter is verified by a simulation example.

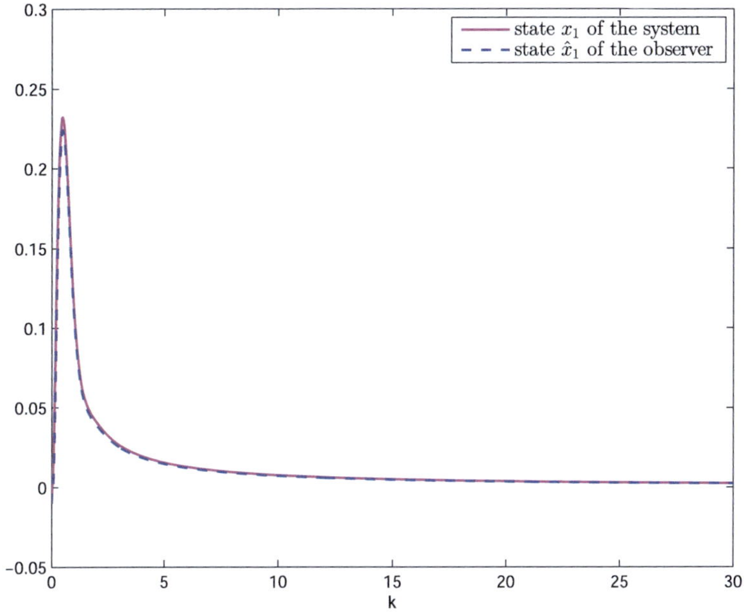

Fig. 2.3 State x_1 and estimation of state x_1

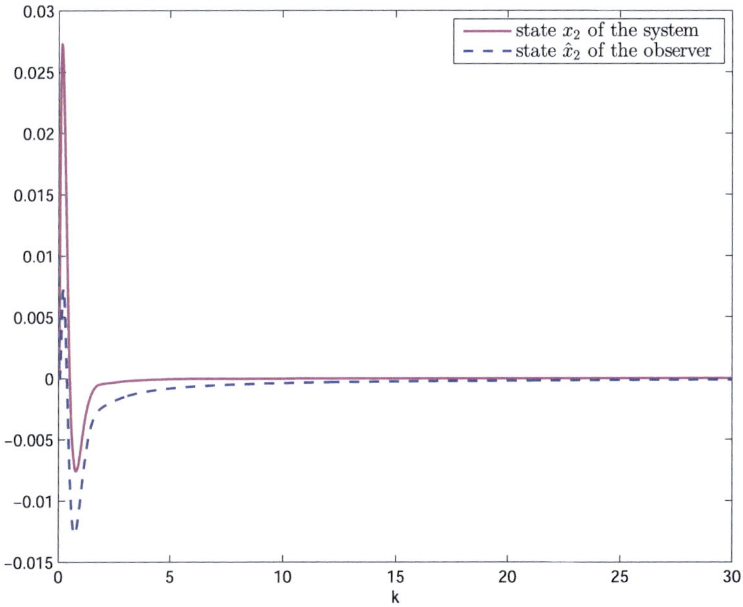

Fig. 2.4 State x_2 and the estimation of state x_2

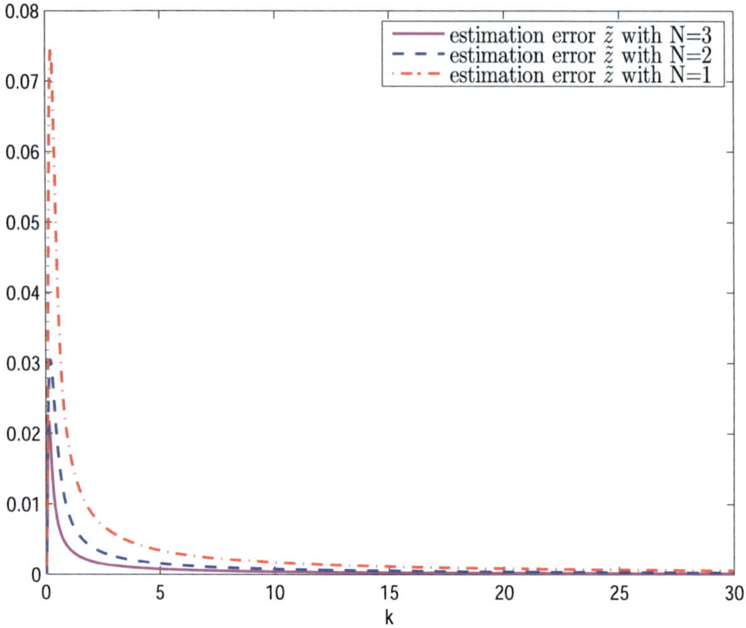

Fig. 2.5 Estimation errors $\tilde{z}$ with $N = 3$, $N = 2$ and $N = 1$ (the method of [11])

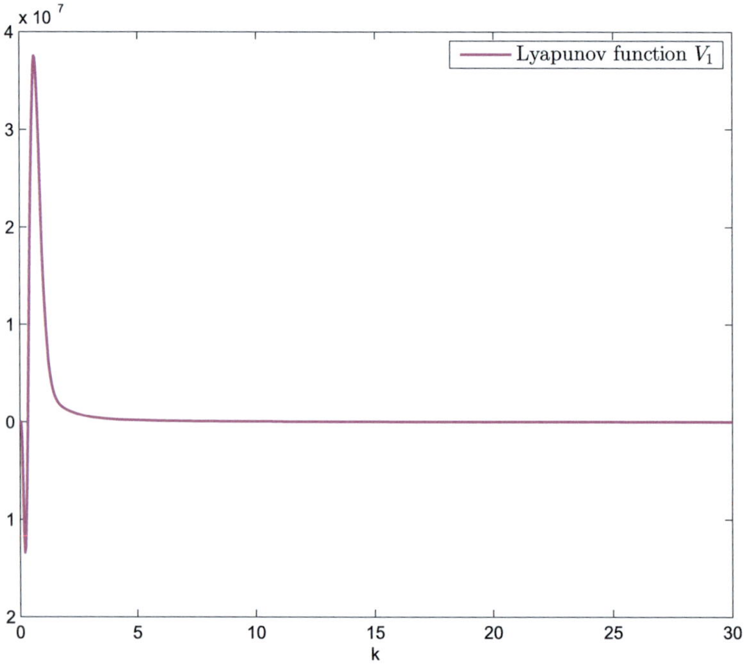

Fig. 2.6 Lyapunov function V_1

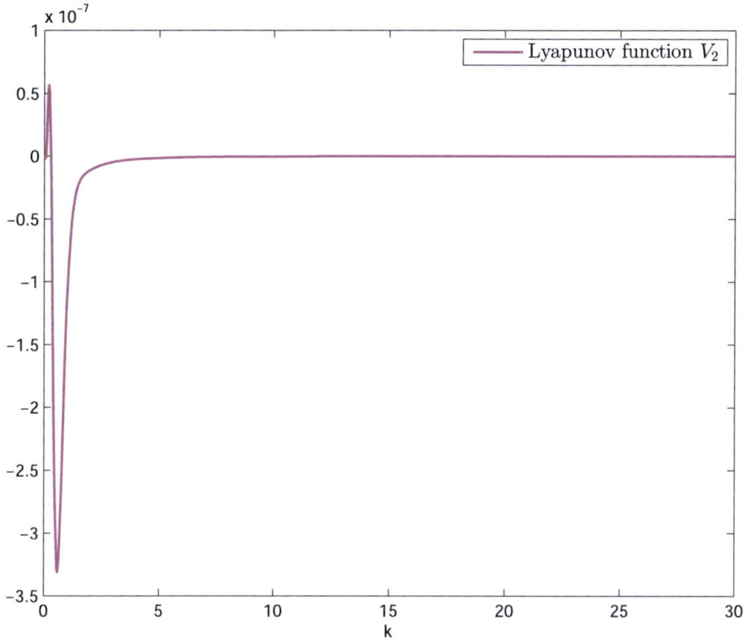

Fig. 2.7 Lyapunov function V_2

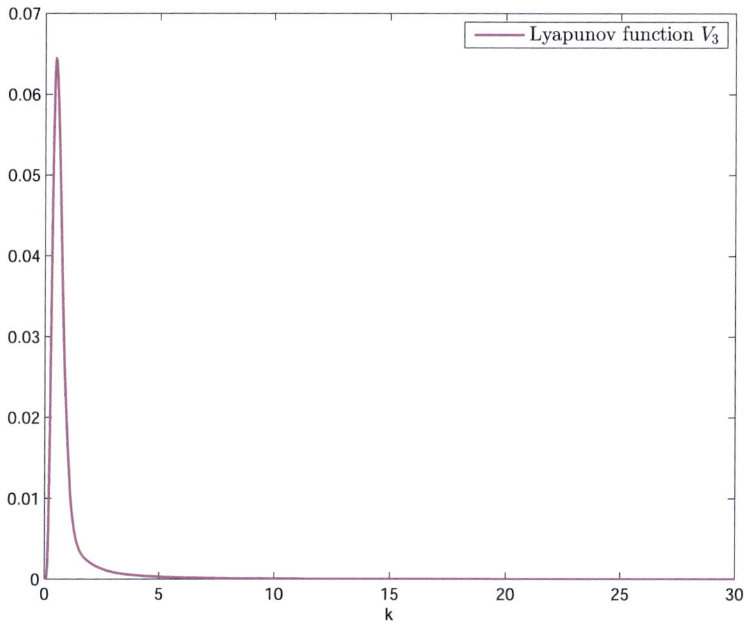

Fig. 2.8 Lyapunov function V_3

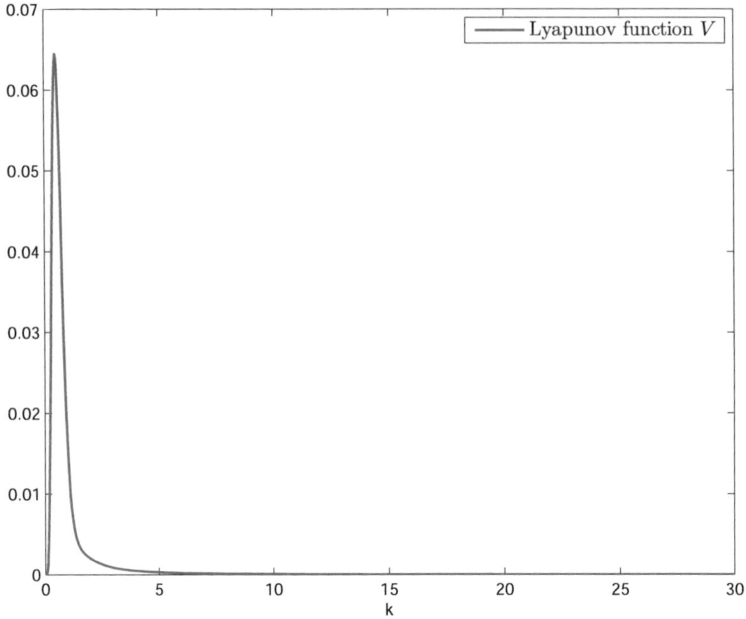

Fig. 2.9 Lyapunov function V

2.6 Notes

The Lyapunov function analysis method is considered a powerful tool to ensure the stability of the systems [1, 2], therefore, it is necessary to explore new Lyapunov functions to reduce the conservatism of the existing filtering results. Meanwhile, the states in practical systems are not always measurable, which poses challenges to the design of filters. Therefore, it is very important to study the H_∞ filtering strategy of T-S fuzzy systems with immeasurable state variables by proposing a new structured Lyapunov function, which motivates this chapter. In addition, it should be noted that the results presented in this chapter are mainly constructed based on the ones in Ref. [23].

References

1. Li HY, Gao YB, Shi P, Zhao XD (2015) Output-feedback control for T-S fuzzy delta operator systems with time-varying delays via an input-output approach. IEEE Trans Fuzzy Syst 23(4):1100–1112
2. Li XJ, Yang GH (2014) Fault detection for T-S fuzzy systems with unknown membership functions. IEEE Trans Fuzzy Syst 22(1):139–152
3. Su XJ, Wu LG, Shi P (2013) Sensor networks with random link failures: distributed filtering for T-S fuzzy systems. IEEE Trans Indus Inform 9(3):1739–1750

4. Tao J, Lu RQ, Su HY, Shi P, Wu ZG (2018) Asynchronous filtering of nonlinear Markov jump systems with randomly occurred quantization via T-S fuzzy models. IEEE Trans Fuzzy Syst 26(4):1866–1877
5. Yang DS, Xie XP (2017) Observer-based state estimation of discrete-time nonlinear systems via a novel maximum-priority-based fuzzy observer. Signal Process 137:63–68
6. Wang LK, Lam HK (2019) New stability criterion for continuous-time Takagi-Sugeno fuzzy systems with time-varying delay. IEEE Trans Cybern 49:1551–1556
7. Yang DS, Xie XP (2016) Relaxed H_∞ control design of discrete-time Takagi-Sugeno fuzzy systems: A multi-samples approach. Neurocomputing 171(1):106–112
8. Wang LK, Liu XD, Zhang HG (2018) Further studies on H_∞ observer design for continuous-time Takagi-Sugeno fuzzy model. Inform Sci 422:396–407
9. Ye D, Diao NN, Zhao XG (2018) Fault-tolerant controller design for general polynomial-fuzzy-model-based systems. IEEE Trans Fuzzy Syst 26(2):1046–1051
10. Lam HK, Seneviratne LD (2011) Stability analysis of polynomial fuzzy-model-based control systems under perfect/imperfect premise matching. IET Control Theory Appl 5(15):1689–1697
11. Wang HM, Yang GH (2016) Fault detection and isolation for affine fuzzy systems with sensor faults. IEEE Trans Fuzzy Syst 24(5):1058–1071
12. Peng S, Su XJ, Li F (2016) Dissipativity-based filtering for fuzzy switched systems with stochastic perturbation. IEEE Trans Autom Control 61(6):1694–1699
13. Qiu JB, Feng G, Gao HJ (2010) Fuzzy-model-based piecewise H_∞ static-output-feedback controller design for networked nonlinear systems. IEEE Trans Fuzzy Syst 18(5):919–934
14. Ahmadi AA, Parrilo PA (2008) Non-monotonic Lyapunov functions for stability of discrete time nonlinear and switched systems. In: 2008 47th IEEE conference on decision and control, pp 614–621
15. Derakhshan SF, Fatehi A, Sharabiany MG (2014) Nonmonotonic observer-based fuzzy controller designs for discrete time T-S fuzzy systems via LMI. IEEE Trans Cybern 44(12):2557–2567
16. Nasiri A, Nguang SK, Swain A, Almakhles DJ (2018) Reducing conservatism in an H_∞ robust state-feedback control design of T-S fuzzy systems: a nonmonotonic approach. IEEE Trans Fuzzy Syst 26(1):386–390
17. Lacerda MJ, Seiler P (2017) Stability of uncertain systems using Lyapunov functions with non-monotonic terms. Automatica 82:187–193
18. Pan YN, Yang GH (2018) Event-triggered fault detection filter design for nonlinear networked systems. IEEE Trans Syst Man Cybern 48(11):1851–1862
19. Wang XL, Yang GH (2019) Event-triggered H_∞ filtering for discrete-time T-S fuzzy systems via network delay optimization technique. IEEE Trans Syst Man Cybern 49(10):2026–2035
20. Lam HK, Li HL, Deters C, Secco EL, Wurdemann HA, Althoefer K (2014) Control design for interval type-2 fuzzy systems under imperfect premise matching. IEEE Trans Indus Electron 61(2):956–968
21. Wang L, Feng G (2004) Piecewise H_∞ controller design of discrete-time fuzzy systems. IEEE Trans Syst Man Cybern, Part B (Cybern) 34(1):682–686
22. Gao HJ, Zhao Y, Lam J, Chen K (2009) H_∞ fuzzy filtering of nonlinear systems with intermittent measurements. IEEE Trans Fuzzy Syst 17(2):291–300
23. Wang XL, Yang GH (2020) H_∞ observer design for fuzzy system with immeasurable state variables via a new Lyapunov function. IEEE Trans Fuzzy Syst 28(2):236–245

Chapter 3
H_∞ Filtering for T-S Fuzzy Systems With Multiple Time-Varying Delays

3.1 Introduction

In this chapter, the design of piecewise fuzzy H_∞ filters for discrete-time T-S fuzzy systems has been studied without considering the effect of system delay on system performance. However, it should be noted that network delay is usually considered to be one of the main causes of the performance degradation, oscillation or even instability of the engineering systems [1–3]. Therefore, it is necessary to conduct research on the systems with delay, and in the past few decades, scholars have invested extensive attention and achieved fruitful results in addressing delay issues, such as Refs. [4–8]. Moreover, it should be noted that there are two main ways to deal with delay items in the existing research results, one is the delay-independent method [9–11], and the other is delay-dependent method [12–14]. The former is independent of the delay size, and the latter includes the delay size information. Obviously the former is more conservative.

Moreover, it should be pointed out that the results mentioned above are all focus on the single delay problems, and a large number of studies have also been carried out on the problem of multiple delays. The multiple delay problems in Refs. [15, 16] are solved by extending the single delay method to multiple delays. Unfortunately, this simple extension will lead to conservatism. Subsequently, a new Lyapunov function in Ref. [17] is constructed to deal with the problem of multiple constant delays, in which the relationship between two constant delays among multiple constant delays is considered. However, it needs to be pointed out that the Lyapunov function in Ref. [17] cannot be apply to the case of multiple time-varying delays. In Ref. [18], the problem of multiple time-varying delays for T-S fuzzy model has been studied. However, the method proposed in Ref. [18] only uses the current delay information and the same delay upper and lower bounds for the different time-varying delays. Therefore, it may be some conservatism.

In recent years, the multiple delay problems have been addressed by some new techniques proposed by researchers. A delays-dependent region partitioning

X.-L. Wang et al., *Robust Filtering and Fault Detection for T-S Fuzzy Systems*,
https://doi.org/10.1007/978-981-96-5818-3_3

approach is constructed in Ref. [19] to obtain a stability criterion of the linear systems, and furthermore, the upper bound of delay is optimized. However, it should be noted that the current states, the exactly delayed states and the marginally delayed states for adjacent time-varying delays is taken into account in Ref. [19]. In fact, this method does not make full use of information about delay related items. To sum up, in the aforementioned Refs. [18, 19], the method designed in Ref. [18] uses the current delay information and the same delay upper and lower bounds for the different time-varying delays, and the proposed method in Ref. [19] only considers the information of the relationship between the partial delay related items and the current system states. Therefore, there is conservatism in the existing methods, and how to exploit more information of the relationship between the delay related items and the current system states to further obtain a greater delay upper bound will be an important and meaningful problem, which motivates this chapter.

This chapter investigates the problem of H_∞ filter design for discrete-time T-S fuzzy systems with multiple time-varying delays. An improved delays-dependent region partitioning method that takes advantage of all possible information about the relationship between all the delay related terms and the current system states is first proposed. Then, based on this method, a novel Lyapunov functional is established. Subsequently, new H_∞ performance criteria of fuzzy system are obtained by introducing the reciprocally convex method and the Jensen's inequality to deal with the related items resulted from the region partitioning. Moreover, the results obtained with an improved delays-dependent region partitioning method proposed in this chapter can be reduced to the existing results with the non-partition method. Finally, an example is given to verify the effectiveness and superiority of the theory.

3.2 System Modeling and Problem Statement

A discrete-time T-S fuzzy system with multiple time-varying delays is considered, as shown below:

$$\begin{aligned} x(k+1) &= A_h x(k) + \sum_{m=1}^{N} G_{hm} x(k-\tau_m(k)) + B_h w(k) \\ y(k) &= C_h x(k) + D_h w(k) \\ z(k) &= E_h x(k) \end{aligned} \tag{3.1}$$

with

$$A_h = \sum_{i=1}^{r} h_i(\nu(k))A_i,\ G_{hm} = \sum_{i=1}^{r} h_i(\nu(k))G_{im},\ B_h = \sum_{i=1}^{r} h_i(\nu(k))B_i,$$

$$C_h = \sum_{i=1}^{r} h_i(\nu(k))C_i,\ D_h = \sum_{i=1}^{r} h_i(\nu(k))D_i,\ E_h = \sum_{i=1}^{r} h_i(\nu(k))E_i$$

where r is the number of inference rules of the system model. $\nu(k)$ denotes the premise variable that depend on the states of the system model. $x(k) \in R^n$ is the state of the system model. $w(k) \in R^n$ is the disturbance with bounded energy. $y(k) \in R^m$ and $z(k) \in R^m$ are the output vector and the performance output, respectively. A_i, G_{im}, B_i, C_i, D_i and E_i are known constant matrices with appropriate dimensions. $\tau_m(k)$ represents the time-varying delay and satisfies:

$$0 \leq \tau_m(k) \leq \tau_m^M \tag{3.2}$$

where τ_m^M denotes the upper bound of delay, and without losing generality, suppose $\tau_1^M \leq \tau_2^M \leq \cdots \leq \tau_N^M$. Hereinafter, in order to facilitate the theoretical derivation, τ_m is used instead of $\tau_m(k)$ for all $m = 1, 2, \ldots, N$.

Moreover, the membership function $h_i(\nu(k))$ satisfies the following relationships:

$$h_i(\nu(k)) \geq 0, \sum_{i=1}^{r} h_i(\nu(k)) = 1, i = 1, 2, \ldots, r \tag{3.3}$$

The performance output $z(k)$ in T-S fuzzy system (3.1) is estimated by the filter, which is established as follows:

$$\begin{aligned} x_f(k+1) &= A_{fh}x_f(k) + B_{fh}y(k) \\ z_f(k) &= C_{fh}x_f(k) \end{aligned} \tag{3.4}$$

with

$$A_{fh} = \sum_{i=1}^{r} h_i(\nu(k))A_{fi}, B_{fh} = \sum_{i=1}^{r} h_i(\nu(k))B_{fi}, C_{fh} = \sum_{i=1}^{r} h_i(\nu(k))C_{fi}$$

where $x_f(k) \in R^n$ is the state of the filter, and $z_f(k) \in R^m$ is the estimation of the system performance output. A_{fi}, B_{fi} and C_{fi} are the fuzzy filter gains to be designed.

Define $\xi(k) = \left[x^T(k) \; x_f^T(k) \right]^T$ and $\bar{r}(k) = z(k) - z_f(k)$, and then, the structure of the filtering error system can be established from (3.1) and (3.4), as shown below:

$$\begin{aligned} \xi(k+1) &= \bar{A}\xi(k) + \sum_{m=1}^{N} \bar{G}_m\xi(k-\tau_m) + \bar{E}w(k) \\ \bar{r}(k) &= \bar{C}\xi(k) \end{aligned} \tag{3.5}$$

where

$$\bar{A} = \begin{bmatrix} A_h & 0 \\ B_{fh}C_h & A_{fh} \end{bmatrix}, \bar{G}_m = \begin{bmatrix} G_{hm} & 0 \\ 0 & 0 \end{bmatrix}, \bar{E} = \begin{bmatrix} B_h \\ B_{fh}D_h \end{bmatrix}, \bar{C} = \begin{bmatrix} E_h & -C_{fh} \end{bmatrix}$$

Based on the above analysis, the purpose of this study is summed up as the following problems.

Problem: The purpose of this paper is to design a fuzzy filter (3.4) for the T-S fuzzy system (3.1) to estimate the system performance output $z(k)$, and then the filtering error system (3.5) needs to satisfy the following two performance conditions:

(S.1) *Asymptotic Stability*: The filtering error system (3.5) with $w(k) = 0$ is asymptotic stable.

(S.2) H_∞ *Performance*: The filtering error system (3.5) satisfies the following inequality

$$\|\bar{r}(k)\|_2^2 \leq \beta^2 \|w(k)\|_2^2 \tag{3.6}$$

for all $w(k) \in L_2[0, \infty)$, in which β is the H_∞ performance bound.

3.3 H_∞ Fuzzy Filter Design

In this section, in order to further reduce the conservatism and obtain a larger delay upper bound for T-S fuzzy systems with multiple time-varying delay, an improved delays-dependent region partitioning method will be presented. The main ideas of this method can be summarized as: All possible information about the relationship between all the delay related terms (including the Nth delay related terms, the $(N-1)$th delay related terms, ..., the 1th delay related terms in system (3.1)) and the current system states are considered for regional segmentation, and then, based on these segmented regions, a new Lyapunov function for the system (3.1) is established. For example, when the mth delay related terms and the $(m-1)$th delay related terms are considered, the delay interval $[0, \tau_m^M]$ is decomposed into two subintervals $[0, \tau_m]$ and $[\tau_m, \tau_m^M]$, and then one can divide the two subintervals $[0, \tau_m]$ and $[\tau_m, \tau_m^M]$ into four subintervals due to the existence of new variables τ_{m-1} and τ_m^M. These subintervals are grouped into three regional types as shown below:

(1) Type 1: $0 < \tau_{m-1} < \tau_{m-1}^M < \tau_m < \tau_m^M$.
(2) Type 2: $0 < \tau_{m-1} < \tau_m < \tau_{m-1}^M < \tau_m^M$.
(3) Type 3: $0 < \tau_m < \tau_{m-1} < \tau_{m-1}^M < \tau_m^M$.

Then, a new Lyapunov function, which includes the three regional types, are established to ensure the desired performance of the filtering error system.

Next, two useful lemmas, Jensen's inequality and reciprocally convex method, are introduced.

Lemma 3.1 ([20]) *Given a positive constant matrix $Q \in \mathbb{R}^{n\times n}$, any function $x(k)$ in $[a, b] \longrightarrow \mathbb{R}^n$, the following inequality is satisfied:*

$$\sum_{u=a}^{b}(x(u+1)-x(u))^T Q(x(u+1)-x(u))du$$
$$\geq \frac{1}{b-a}(x(b)-x(a))^T Q(x(b)-x(a)) \tag{3.7}$$

Lemma 3.2 ([21, 22]) *Let $\psi_1, \psi_2, \ldots, \psi_N$: $\mathbb{R} \rightarrow R$ be symmetric positive definite matrices in an open subset D of $\mathbb{R}^m$. Then, the reciprocally convex combination of ψ_i over D satisfies:*

$$\min_{\{\alpha_i|\alpha_i>0,\sum_i \alpha_i=1\}} \sum_i \frac{1}{\alpha_i}\psi_i(k) = \sum_i \psi_i(k) + \max_{\{\phi_{i,j}(k)\}} \sum_{i\neq j} \phi_{i,j}(k) \tag{3.8}$$

such that

$$\left\{\phi_{i,j} : \mathbb{R}^m \rightarrow \mathbb{R}, \phi_{j,i}(k) = \phi_{i,j}(k), \begin{bmatrix} \psi_i(k) & \phi_{i,j}(k) \\ \phi_{j,i}(k) & \psi_j(k) \end{bmatrix} \geq 0\right\}.$$

3.3.1 H_∞ Performance Analysis Conditions

Now, the improved delays-dependent region partitioning method is used to further reduce the conservatism and obtain a larger delay upper bound for multiple time-varying delay systems. Based on the proposed method, new H_∞ performance criteria of multiple time-varying delay systems are obtained by introducing the reciprocally convex method and the Jensen's inequality to deal with the new Lyapunov functional. Before presenting the theorem, define $a_{m,0}, a_{m,1}, a_{m,2}, \ldots, a_{m,2m-1}$ and $a_{m,2m}$ as the ascending sorted values of 0, τ_1, τ_1^M, τ_2, τ_2^M, …, τ_m and τ_m^M for $2 \leq m \leq N$.

Theorem 3.1 *For the T-S fuzzy system (3.1), the filtering error system (3.5) satisfies the performance condition (S.2), if there exist matrices $P > 0$, $Q_{m,q} > 0$, $N_m > 0$, $R_{m,q} > 0$, $X_{1,m}$, $X_{2,m}$, …, $X_{m-1,m-1}$ and $\beta > 0$ such that the following inequality is satisfied for τ_m in $[0, \tau_m^M]$, $(1 \leq m \leq N)$, $(1 \leq q \leq 2m)$*

$$\Phi(k) < 0 \tag{3.9}$$

where

$$\begin{aligned}\Phi(k) = {} & \Gamma_1^T P\Gamma_1 - I_1^T P I_1 + \sum_{m=1}^{N}(I_1^T Q_{m,1} I_1 + I_2^T(Q_{m,2} - Q_{m,1})I_2 \\ & + I_3^T(Q_{m,3} - Q_{m,2})I_3 \\ & + \cdots + I_{2N-1}^T(Q_{m,2m} - Q_{m,2m-1})I_{2N} - I_{2N}^T Q_{m,2m} I_{2N}) \\ & + \sum_{m=1}^{N}(\tau_m^M(\tau_1^M - \tau_0^M)\Gamma_2^T R_{m,1}\Gamma_2 + \tau_m^M(\tau_2^M - \tau_1^M)\Gamma_2^T R_{m,2}\Gamma_2\end{aligned}$$

$$
\begin{aligned}
&+\cdots+\tau_m^M(\tau_m^M-\tau_{m-1}^M)\Gamma_2^T R_{m,m}\Gamma_2) \\
&-\sum_{m=1}^{N}\begin{bmatrix} I_1-\bar{H}_{x_1} \\ \bar{H}_{x_1}-\bar{H}_{x_2} \\ \vdots \\ \bar{H}_{x_{m-1}}-\bar{H}_{x_m}\end{bmatrix}^T \begin{bmatrix} R_{m,1} & X_{1,2} & \cdots & X_{1,m} \\ \star & R_{m,2} & \cdots & X_{2,m} \\ \star & \star & \ddots & \vdots \\ \star & \star & \star & R_{m,m}\end{bmatrix}\begin{bmatrix} I_1-\bar{H}_{x_1} \\ \bar{H}_{x_1}-\bar{H}_{x_2} \\ \vdots \\ \bar{H}_{x_{m-1}}-\bar{H}_{x_m}\end{bmatrix} \\
&+\sum_{m=1}^{N}(I_1^T N_m I_1 - H_{x_m}^T N_m H_{x_m}) + (\bar{C}I_1)^T(\bar{C}I_1) - \beta^2 I_{2N+2}^T I_{2N+2}
\end{aligned}
$$

in which

$$
\Gamma_1 = \bar{A}I_1 + \sum_{m=1}^{N}\bar{G}_m H_{xm} + \bar{E}I_{2N+2},
$$

$$
\Gamma_2 = \bar{B}I_1 + \sum_{m=1}^{N}\bar{G}_m H_{m+1} + \bar{E}I_{2N+2}
$$

and

$$
I_1 = \Big[\, I\ 0\ 0\ \underbrace{\cdots}_{N-3}\ 0\ 0\ 0\ \underbrace{\cdots}_{N-3}\ 0\ 0 \,\Big],
$$

$$
\vdots
$$

$$
I_{2N+2} = \Big[\, 0\ 0\ 0\ \underbrace{\cdots}_{N-3}\ 0\ 0\ 0\ \underbrace{\cdots}_{N-3}\ 0\ I \,\Big]
$$

Proof A suitable Lyapunov function for the filtering error system (3.5) is constructed and presented as follows:

$$
\mathbb{V}(k) = \mathbb{V}_1(k) + \mathbb{V}_2(k) + \mathbb{V}_3(k) + \mathbb{V}_4(k) \tag{3.10}
$$

and

$$
\mathbb{V}_1(k) = \xi(k)^T P \xi(k) \tag{3.11}
$$

$$
\mathbb{V}_2(k) = \sum_{m=1}^{N}\sum_{q=1}^{2m}\sum_{v=k-a_{m,q}}^{k-a_{m,q-1}-1}\xi(v)^T Q_{m,q}\xi(v) \tag{3.12}
$$

$$
\mathbb{V}_3(k) = \sum_{m=1}^{N}\sum_{v=k-\tau_m}^{k-1}\xi(v)^T N_m \xi(v) \tag{3.13}
$$

$$
\mathbb{V}_4(k) = \sum_{m=1}^{N}\tau_m^M\sum_{q=1}^{N}\sum_{v=-\tau_q^M}^{-\tau_{q-1}^M-1}\sum_{\rho=k+v}^{k-1}\varphi^T(\rho)R_{m,q}\varphi(\rho) \tag{3.14}
$$

in which $\tau_0^M = 0$, and

$$\begin{aligned}\varphi(k) &= \xi(k+1) - \xi(k) \\ &= \bar{B}\xi(k) + \sum_{m=1}^{N} \bar{G}_m \xi(k-\tau_m) + \bar{E}w(k)\end{aligned}$$

with $\bar{B} = (\bar{A} - I)$.

In addition, $P > 0$, $Q_{m,q} > 0$, $N_m > 0$ and $R_{m,q} > 0$ are matrices to be designed.

Denote $\Delta\mathbb{V}(k) = \mathbb{V}(k+1) - \mathbb{V}(k)$, and then one can obtain the following equation relations from the filtering error system (3.5):

$$\begin{aligned}\Delta\mathbb{V}_1(k) &= \mathbb{V}_1(k+1) - \mathbb{V}_1(k) \\ &= \xi(k+1)^T P\xi(k+1) - \xi(k)^T P\xi(k)\end{aligned} \tag{3.15}$$

$$\begin{aligned}\Delta\mathbb{V}_2(k) &= \mathbb{V}_2(k+1) - \mathbb{V}_2(k) \\ &= \sum_{m=1}^{N}(\xi(k-a_{m,0})^T Q_{m,1}\xi(k-a_{m,0}) - \xi(k-a_{m,1})^T Q_{m,1}\xi(k-a_{m,1}) \\ &\quad + \xi(k-a_{m,1})^T Q_{m,2}\xi(k-a_{m,1}) - \xi(k-a_{m,2})^T Q_{m,2}\xi(k-a_{m,2}) + \cdots \\ &\quad + \xi(k-a_{m,2m-2})^T Q_{m,2m-1}\xi(k-a_{m,2m-2}) \\ &\quad - \xi(k-a_{m,2m-1})^T Q_{m,2m-1}\xi(k-a_{m,2m-1}) \\ &\quad + \xi(k-a_{m,2m-1})^T Q_{m,2m}\xi(k-a_{m,2m-1}) \\ &\quad - \xi(k-a_{m,2m})^T Q_{m,2m}\xi(k-a_{m,2m}))\end{aligned} \tag{3.16}$$

$$\begin{aligned}\Delta\mathbb{V}_3(k) &= \mathbb{V}_3(k+1) - \mathbb{V}_3(k) \\ &= \sum_{m=1}^{N}(\xi(k)^T N_m\xi(k) - \xi(k-\tau_m)^T N_m\xi(k-\tau_m))\end{aligned} \tag{3.17}$$

$$\begin{aligned}\Delta\mathbb{V}_4(k) &= \mathbb{V}_4(k+1) - \mathbb{V}_4(k) \\ &= \sum_{m=1}^{N} \tau_m^M((\tau_1^M - \tau_0^M)\varphi^T(k)R_{m,1}\varphi(k) - \sum_{v=k-\tau_1^M}^{k-\tau_0^M-1} \varphi^T(v)R_{m,1}\varphi(v)\end{aligned} \tag{3.18}$$

$$+ (\tau_2^M - \tau_1^M)\varphi^T(k)R_{m,2}\varphi(k) - \sum_{v=k-\tau_2^M}^{k-\tau_1^M-1} \varphi^T(v)R_{m,2}\varphi(v) + \cdots$$

$$+ (\tau_m^M - \tau_{m-1}^M)\varphi^T(k)R_{m,m}\varphi(k) - \sum_{v=k-\tau_m^M}^{k-\tau_{m-1}^M-1} \varphi^T(v)R_{m,m}\varphi(v))$$

For $\Delta\mathbb{V}_4(k)$, by Lemmas 3.1 and 3.2, one has:

$$\begin{aligned}
&-\tau_m^M \sum_{v=k-\tau_1^M}^{k-\tau_0^M-1} \varphi^T(v)R_{m,1}\varphi(v) - \tau_m^M \sum_{v=k-\tau_2^M}^{k-\tau_1^M-1} \varphi^T(v)R_{m,2}\varphi(v) \\
&-\cdots\tau_m^M \sum_{v=k-\tau_m^M}^{k-\tau_{m-1}^M-1} \varphi^T(v)R_{m,m}\varphi(v) \\
\le& -\frac{\tau_m^M}{\tau_1^M - \tau_0^M}(\varphi(k-\tau_0^M) - \varphi(k-\tau_1^M)^T R_{m,1}(\varphi(k-\tau_0^M) - \varphi(k-\tau_1^M)) \\
&-\cdots - \frac{\tau_m^M}{\tau_m^M - \tau_{m-1}^M}(\varphi(k-\tau_{m-1}^M) - \varphi(k-\tau_m^M)^T R_{m,1}(\varphi(k-\tau_{m-1}^M) - \varphi(k-\tau_m^M)) \\
\le& \begin{bmatrix} \xi(k-\tau_0^M) - \xi(k-\tau_1^M) \\ \xi(k-\tau_1^M) - \xi(k-\tau_2^M) \\ \vdots \\ \xi(k-\tau_{m-1}^M) - \xi(k-\tau_m^M) \end{bmatrix}^T \begin{bmatrix} R_{m,1} & X_{1,2} & \cdots & X_{1,m} \\ \star & R_{m,2} & \cdots & X_{2,m} \\ \star & \star & \ddots & \vdots \\ \star & \star & \star & R_{m,m} \end{bmatrix} \\
\times& \begin{bmatrix} \xi(k-\tau_0^M) - \xi(k-\tau_1^M) \\ \xi(k-\tau_1^M) - \xi(k-\tau_2^M) \\ \vdots \\ \xi(k-\tau_{m-1}^M) - \xi(k-\tau_m^M) \end{bmatrix}
\end{aligned} \tag{3.19}$$

Define the following equation:

$$J(k) = \Delta\mathbb{V}(k) + \bar{r}^T(k)\bar{r}(k) - \beta^2 w^T(k)w(k) \tag{3.20}$$

and

$$\begin{aligned}
\varepsilon(k) = \Big[& \xi^T(k)\ \xi^T(k-a_{m,0})\ \xi^T(k-a_{m,1})\ \underbrace{\cdots}_{N-3}\ \xi^T(k-a_{m,m})\ \xi^T(k-a_{m,m+1}) \\
& \xi^T(k-a_{m,m+2})\ \underbrace{\cdots}_{N-3}\ \xi^T(k-a_{m,2m})\ w^T(k)\Big]^T
\end{aligned}$$

and

$$H_{x_m}\varepsilon(k) = \xi(k-\tau_m),\ \bar{H}_{x_m}\varepsilon(k) = \xi(k-\tau_m^M),\ x_m \in [2, 2N+1]$$

Then, it is known from (3.15)–(3.20) that the following inequality is satisfied:

$$\begin{aligned} J(k) &= \Delta\mathbb{V}(k) + \bar{r}^T(k)\bar{r}(k) - \beta^2 w^T(k)w(k) \\ &\le \varepsilon^T(k)\Phi(k)\varepsilon(k) \end{aligned} \tag{3.21}$$

According to (3.9), the inequality $J(k) = \Delta\mathbb{V}(k) + \bar{r}^T(k)\bar{r}(k) - \beta^2 w^T(k)w(k) < 0$ holds.

Based on the above analysis, it is concluded that the asymptotical stability of the filtering error system (3.5) with H_∞ performance is ensured. The proof is completed. □

Remark 3.1 In this section, in order to obtain the H_∞ performance criterion of the system (3.1), a novel Lyapunov functional, which includes all the segmented regions, is first established. Then, Jensen's inequality and reciprocally convex method are used to the terms resulted from the region partitioning.

The results obtained with an improved delays-dependent region partitioning method in this chapter can be reduced to the existing result in Ref. [18] by using the non-partition method and replacing the matrices $Q_{m,q}$, N_m and $R_{m,q}$ as Q, N and R, simultaneously. Moreover, the result obtained by using the existing method in Ref. [18] is summed up as the following corollary:

Corollary 3.1 *For the T-S fuzzy system (3.1), the filtering error system (3.5) satisfies the performance condition (S.2), if there exist matrices $P > 0$, $Q_m > 0$, $N_m > 0$, $R_m > 0$ and $\beta > 0$ such that the following inequality is satisfied for τ_m in $[0, \tau_m^M]$, $(1 \le m \le N)$*

$$\bar{\Phi}(k) < 0 \tag{3.22}$$

where

$$\begin{aligned} \bar{\Phi}(k) = {} & \Gamma_3^T P \Gamma_3 - I_1^T P I_1 + I_1^T Q_1 I_1 - I_4^T Q_1 I_4 + I_1^T Q_2 I_1 - I_5^T Q_2 I_5 + I_1^T N_1 I_1 \\ & - I_2^T N_1 I_2 + I_1^T N_2 I_1 - I_3^T N_2 I_3 + (\tau_1^M)^2 \Gamma_4^T R_1 \Gamma_4 + (\tau_2^M)^2 \Gamma_4^T R_2 \Gamma_4 \\ & - (I_4 - I_1)^T R_1 (I_4 - I_1) - (I_5 - I_1)^T R_2 (I_5 - I_1) + \bar{C}^T \bar{C} - \beta^2 w^T(k) w(k) \end{aligned}$$

and

$$\begin{aligned} \Gamma_3 &= \bar{A} I_1 + \bar{G}_1 I_2 + \bar{G}_2 I_3 + \bar{E} I_6, \\ \Gamma_4 &= \bar{B} I_1 + \bar{G}_1 I_2 + \bar{G}_2 I_3 + \bar{E} I_6 \end{aligned}$$

Proof A suitable Lyapunov function for the filtering error system (3.5) is constructed and presented as follows:

$$\mathbb{V}(k) = \mathbb{V}_1(k) + \mathbb{V}_2(k) + \mathbb{V}_3(k) + \mathbb{V}_4(k) \tag{3.23}$$

and

$$\mathbb{V}_1(k) = \xi(k)^T P\xi(k) \tag{3.24}$$

$$\mathbb{V}_2(k) = \sum_{m=1}^{N} \sum_{v=k-\tau_m^M}^{k-1} \xi(v)^T Q_m \xi(v) \tag{3.25}$$

$$\mathbb{V}_3(k) = \sum_{m=1}^{N} \sum_{v=k-\tau_m}^{k-1} \xi(v)^T N_m \xi(v) \tag{3.26}$$

$$\mathbb{V}_4(k) = \sum_{m=1}^{N} \tau_m^M \sum_{v=-\tau_m^M}^{-1} \sum_{\rho=k+v}^{k-1} \varphi^T(\rho) R_m \varphi(\rho) \tag{3.27}$$

in which $\tau_0^M = 0$, and $\varphi(k)$ has been defined in Theorem 3.1.

Denote $\Delta\mathbb{V}(k) = \mathbb{V}(k+1) - \mathbb{V}(k)$, and then the following equation relations from the filtering error system (3.5) is obtained:

$$\begin{aligned}
\Delta\mathbb{V}(k) &= \Delta\mathbb{V}_1(k) + \Delta\mathbb{V}_2(k) + \Delta\mathbb{V}_3(k) + \Delta\mathbb{V}_4(k) \\
&= \xi^T(k+1)P\xi(k+1) - \xi^T(k)P\xi(k) \\
&\quad + \sum_{m=1}^{N} (\xi^T(k) Q_m \xi(k) - \xi^T(k-\tau_m^M) Q_m \xi(k-\tau_m^M)) \\
&\quad + \sum_{m=1}^{N} \tau_m^M ((\tau_m^M - 0)\varphi^T(k) R_m \varphi(k) - \sum_{v=k-\tau_m^M}^{k-1} \varphi^T(v) R_m \varphi(v)) \\
&\quad + \sum_{m=1}^{N} (\xi^T(k) N_m \xi(k) - \xi^T(k-\tau_m^M) N_m \xi(k-\tau_m^M))
\end{aligned} \tag{3.28}$$

Also define

$$\bar{J}(k) = \Delta\mathbb{V}(k) + \bar{r}^T(k)\bar{r}(k) - \beta^2 w^T(k)w(k) \tag{3.29}$$

$$\bar{\varepsilon}(k) = \left[\, \xi^T(k)\ \xi^T(k-\tau_1)\ \xi^T(k-\tau_2)\ \xi^T(k-\tau_1^M)\ \xi^T(k-\tau_2^M)\ w^T(k) \,\right]^T \tag{3.30}$$

Then, one has

$$\begin{aligned}
\bar{J}(k) &= \Delta\mathbb{V}(k) + \bar{r}^T(k)\bar{r}(k) - \beta^2 w^T(k)w(k) \\
&\le \xi^T(k+1)P\xi(k+1) - \xi^T(k)P\xi(k) \\
&\quad + \sum_{m=1}^{N} (\xi^T(k) Q_m \xi(k) - \xi^T(k-\tau_m^M) Q_m \xi(k-\tau_m^M))
\end{aligned} \tag{3.31}$$

$$
\begin{aligned}
&+\sum_{m=1}^{N}(\tau_m^M)^2\varphi^T(k)R_m\varphi(k)-\sum_{m=1}^{N}(\xi(k-\tau_m^M)-\xi(k))^T R_m(\xi(k-\tau_m^M)-\xi(k))\\
&+\sum_{m=1}^{N}(\xi^T(k)N_m\xi(k)-\xi^T(k-\tau_m^M)N_m\xi(k-\tau_m^M))+\bar{C}^T\bar{C}-\beta^2w^T(k)w(k)\\
&=\bar{\varepsilon}^T(k)\bar{\Phi}(k)\bar{\varepsilon}(k)
\end{aligned}
$$

According to (3.22), the inequality $\bar{J}(k)=\Delta\mathbb{V}(k)+\bar{r}^T(k)\bar{r}(k)-\beta^2w^T(k)w(k)<0$ holds. The proof is completed. □

Remark 3.2 In Refs. [18, 19], the problems of multiple time-varying delays have been studied. Unfortunately, the methods in Refs. [18, 19] do not take full advantage of the information about the delay related items and only the information of the relationship between the partial delay related items and the current system states is considered. Thus, there is some conservatism. In this chapter, the proposed improved delays-dependent region partitioning method can make full use of all possible information about the relationship between all the delay related terms and the current system states. Thus, the conservatism can be further reduced.

3.3.2 H_∞ Fuzzy Filter Design Conditions

Note that in inequality (3.9), the matrices P, $R_{m,1}$, $R_{m,2}$, … $R_{m,m}$ are coupled with the augmented matrices $\bar{A}$, $\bar{B}$, $\bar{G}_1$, $\bar{G}_2$, …, $\bar{G}_N$ and $\bar{E}$, respectively. Therefore, it cannot be solved directly by linear matrix inequality toolbox. Next, a linearization method for inequality (3.9) is summed up as Theorem 3.2, as shown below.

Theorem 3.2 *For the T-S fuzzy system (3.1), the filtering error system (3.5) satisfies the performance condition (S.2), if there exist matrices $P>0$, $Q_{m,q}>0$, $N_m>0$, $R_{m,q}>0$, $X_{1,m}$, $X_{2,m}$, …, $X_{m-1,m-1}$, $\beta>0$, Z_j, W_j, $Y=\begin{bmatrix} Y_{11} & Y_{12}\\ Y_{22} & Y_{22}\end{bmatrix}$, in which $Y_{22}>0$ such that the following inequality is satisfied for τ_m in $[0,\tau_m^M]$, $(1\le m\le N)$, $(1\le q\le 2m)$*

$$
\hat{\Phi}(k)<0 \tag{3.32}
$$

where

$$
\begin{aligned}
\hat{\Phi}(k)=\;&\Gamma_1^TY(P-Y-Y^T)Y^T\Gamma_1-I_1^TPI_1+\sum_{m=1}^{N}(I_1^TQ_{m,1}I_1+I_2^T(Q_{m,2}-Q_{m,1})I_2\\
&+I_3^T(Q_{m,3}-Q_{m,2})I_3+\cdots+I_{2N-1}^T(Q_{m,2m}-Q_{m,2m-1})I_{2N}-I_{2N}^TQ_{m,2m}I_{2N})\\
&+\sum_{m=1}^{N}(\tau_m^M(\tau_1^M-\tau_0^M)\Gamma_2^TY(R_{m,1}-Y-Y^T)Y^T\Gamma_2\\
&+\tau_m^M(\tau_2^M-\tau_1^M)\Gamma_2^TY(R_{m,2}-Y-Y^T)Y^T\Gamma_2
\end{aligned}
$$

$$
\begin{aligned}
&+\cdots+\tau_m^M(\tau_m^M-\tau_{m-1}^M)\Gamma_2^T Y(R_{m,m}-Y-Y^T)Y^T\Gamma_2)\\
&-\sum_{m=1}^{N}\begin{bmatrix} I_1-\bar{H}_{x_1}\\ \bar{H}_{x_1}-\bar{H}_{x_2}\\ \vdots \\ \bar{H}_{x_{m-1}}-\bar{H}_{x_m}\end{bmatrix}^T\begin{bmatrix} R_{m,1} & X_{1,2} & \cdots & X_{1,m}\\ \star & R_{m,2} & \cdots & X_{2,m}\\ \star & \star & \ddots & \vdots\\ \star & \star & \star & R_{m,m}\end{bmatrix}\begin{bmatrix} I_1-\bar{H}_{x_1}\\ \bar{H}_{x_1}-\bar{H}_{x_2}\\ \vdots \\ \bar{H}_{x_{m-1}}-\bar{H}_{x_m}\end{bmatrix}\\
&+\sum_{m=1}^{N}(I_1^T N_m I_1-H_{x_m}^T N_m H_{x_m})+(\bar{C}I_1)^T(\bar{C}I_1)-\beta^2 I_{2N+2}^T I_{2N+2}
\end{aligned}
$$

in which the related parameters have been defined in Theorem 3.1.

If the inequality condition (3.32) is feasible, the gain matrices of the filter are obtained as follows:

$$
A_{fj}=(Z_jY_{22}^{-1})^T,\ B_{fj}=(W_jY_{22}^{-1})^T,\ j=1,2,\ldots,r \tag{3.33}
$$

Proof A new matrix variable Y is introduced and defined as $Y=\begin{bmatrix} Y_{11} & Y_{12}\\ Y_{22} & Y_{22}\end{bmatrix}$, in which $Y_{22}>0$. Then, the Schur complement formula is used for inequality (3.9), and the congruence transformation $diag\{\underbrace{I,I,I,\ldots,I,I,\ldots,I,I}_{2N+2},\underbrace{Y,Y,\ldots,Y}_{2N}\}$ is further applied to it. Moreover, define two new matrix variables $Z_j=A_{fj}^T Y_{22}$, $W_j=B_{fj}^T Y_{22}$, and use the inequalities $-Y^TP^{-1}Y\le P-Y-Y^T$, $-Y^TR_{m,1}^{-1}Y\le P-Y-Y^T$, $-Y^TR_{m,2}^{-1}Y\le P-Y-Y^T$, ..., $-Y^TR_{m,m}^{-1}Y\le P-Y-Y^T$. Then, (3.9) is linearized to (3.32). The proof is completed. □

Remark 3.3 The problem of multiple time-varying delays for linear systems is investigated in Ref. [19], and the result is obtained by using the delays-dependent region partitioning approach, in which only the mth delay related terms and the $(m-1)$th delay related terms are considered, which may be some conservatism. In this chapter, in order to further reduce the conservatism and obtain a larger delay upper bound for multiple time-varying delay systems, an improved delays-dependent region partitioning method is proposed, and based on this method, a novel Lyapunov functional is established. Moreover, a new H_∞ performance criterion of T-S fuzzy system is obtained by introducing the reciprocally convex method and the Jensen's inequality to deal with the related items resulted from the region partitioning.

Remark 3.4 Note that the complexity of this method depends heavily on the number of delay related terms. When more delay related terms are considered, more regional types need to be divided, thus more variables and linear matrix inequalities need to be solved. Therefore, there should be a trade-off between increasing the multiple delay upper bound and the computational burdens.

3.4 Numeral Example

In this section, the validity and advantages of the method proposed in this chapter is verified by an example of the tunnel diode circuit in Ref. [23]. The diagram of the tunnel diode circuit is presented in Fig. 2.1 in Chap. 2, the membership functions of the system are shown in Fig. 3.1. Denote $x_1(k) = v_c(k)$ and $x_2(k) = i_L(k)$, the system model is discretized with the sampling period $T_s = 20ms$, and then the following discrete-time T-S fuzzy system is obtained:

$$
\begin{aligned}
x(k+1) &= \sum_{i=1}^{2} h_i(v(k))(A_i x(k) + B_i w(k)) \\
y(k) &= \sum_{i=1}^{2} h_i(v(k))(C_i x(k) + D_i w(k)) \\
z(k) &= \sum_{i=1}^{2} h_i(v(k)) E_i x(k)
\end{aligned}
$$

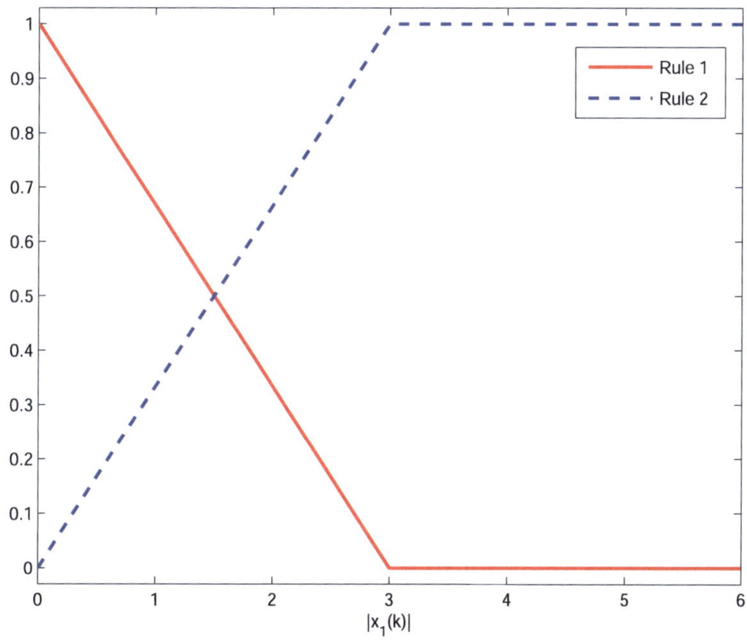

Fig. 3.1 Membership functions

where

$$A_1 = \begin{bmatrix} 0.9887 & 0.9024 \\ -0.0180 & 0.8100 \end{bmatrix}, A_2 = \begin{bmatrix} 0.9034 & 0.8617 \\ -0.0172 & 0.8103 \end{bmatrix}, B_1 = \begin{bmatrix} 0.0093 \\ 0.0181 \end{bmatrix},$$
$$B_2 = \begin{bmatrix} 0.0091 \\ 0.0181 \end{bmatrix}, C_1 = \begin{bmatrix} 1 & 0 \end{bmatrix}, C_2 = \begin{bmatrix} 1 & 0 \end{bmatrix}, D_1 = 1, D_2 = 1,$$
$$E_1 = \begin{bmatrix} 1 & 0 \end{bmatrix}, E_2 = \begin{bmatrix} 1 & 0 \end{bmatrix}$$

In this example, assume $N = 2$, and then the corresponding delay matrices can be denoted as follows:

$$G_{11} = \begin{bmatrix} -0.014 & 0 \\ 0 & -0.078 \end{bmatrix}, G_{12} = \begin{bmatrix} -0.02 & 0 \\ 0 & -0.01 \end{bmatrix},$$
$$G_{21} = \begin{bmatrix} -0.017 & 0 \\ 0 & -0.07 \end{bmatrix}, G_{22} = \begin{bmatrix} -0.019 & 0 \\ 0 & -0.013 \end{bmatrix}$$

In this chapter, the filtering problem for the systems with multiple time-varying delay is studied, and two purposes need to be considered: (1) The filtering error system (3.5) needs to satisfy two performance conditions: asymptotic stability (S.1), H_∞ performance condition (S.2). (2) Obtain the maximum allowable upper bounds for multiple delays. Therefore, the validity and advantages of the proposed method are verified in the following two cases:

Case I: In this case, the advantages of the improved delays-dependent region partitioning method is verified. For this purpose, assume $N = 2$, and therefore the delay interval $[0, \tau_2^M]$ is grouped into three regional types, which are shown as follows:

(1) Type 1: $0 < \tau_1 < \tau_1^M < \tau_2 < \tau_2^M$.
(2) Type 2: $0 < \tau_1 < \tau_2 < \tau_1^M < \tau_2^M$.
(3) Type 3: $0 < \tau_2 < \tau_1 < \tau_1^M < \tau_2^M$.

Table 3.1 lists the maximum delay upper bounds obtained by decomposing the delay interval $[0, \tau_2^M]$ into three regional types and using the existing method in Ref. [18], respectively. It can be seen from Table 3.1 that the upper bounds of multiple delays obtained by using the approach presented in this chapter are greater than the upper bound obtained by using the existing method in Ref. [18]. In addition, the same delay upper bounds are obtained in three different regional types. Therefore, it is concluded that the delay upper bounds obtained by using the improved delays-dependent region partitioning method are larger than the upper bound of delays obtained by using the existing method in Ref. [18]. Thus, the advantages of the proposed method is verified.

Case II: In this case, the effectiveness of the filter design method for multiple time-varying delay systems is verified. For this purpose, assume $N = 2$, $\tau_1^M = 2$, $\tau_2^M = 4$, and the disturbance signal satisfies $w(t) = sin(k)$ $(10 < t < 20)$.

By solving (3.32), the gain matrices of the filter are obtained:

Table 3.1 Maximum allowable delay upper bounds

Methods	Corollary 1	Regional type 1	Regional type 2	Regional type 3
τ_1^M	1	3	3	3
τ_2^M	2	5	5	5

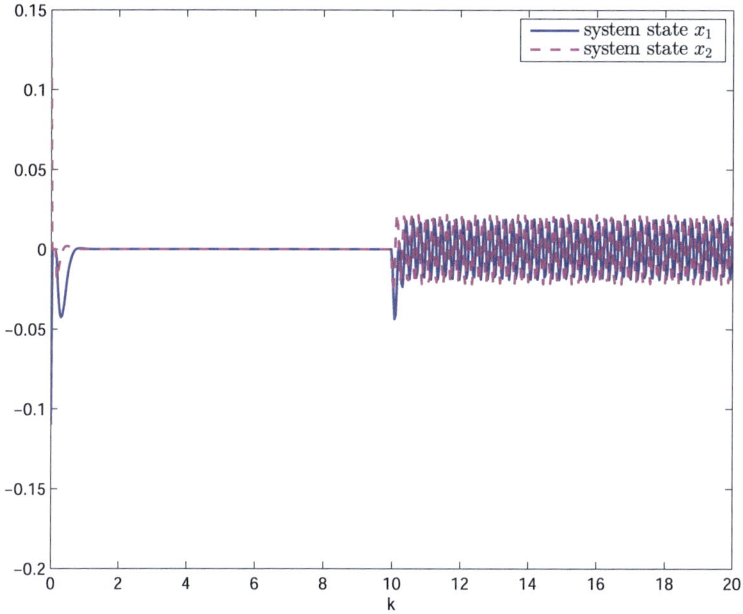

Fig. 3.2 System states x_1 and x_2

$$A_{f1} = \begin{bmatrix} 0.9975 & 0.2551 \\ -0.0061 & 0.9035 \end{bmatrix}, A_{f2} = \begin{bmatrix} 0.9976 & 0.2551 \\ -0.0060 & 0.9034 \end{bmatrix}, B_{f1} = \begin{bmatrix} 0.1210 \\ 0.0188 \end{bmatrix},$$
$$B_{f2} = \begin{bmatrix} 0.1209 \\ 0.0187 \end{bmatrix}, C_{f1} = \begin{bmatrix} -0.9275 & -0.3139 \end{bmatrix}, C_{f2} = \begin{bmatrix} -0.9275 & -0.3138 \end{bmatrix}$$

Based on the obtained gain matrices and the prior given parameters, the system states and filter states are presented in Figs. 3.2 and 3.3, respectively. The system output and filter output are plotted in Fig. 3.4. In Fig. 3.5, the output estimation error is presented, from which it is seen that the method proposed in this chapter can effectively suppress the disturbance signal, and the output estimation error of the system can converge to zero asymptotically without disturbance signal.

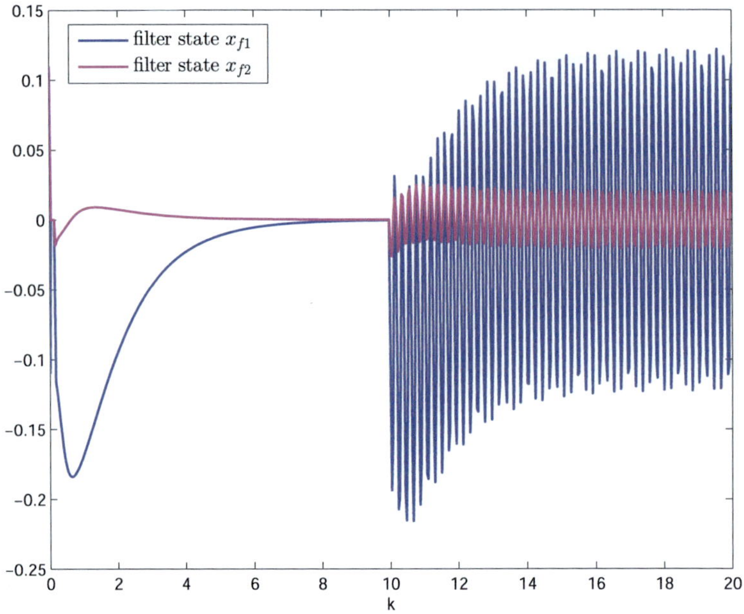

Fig. 3.3 Filter states x_{f1} and x_{f2}

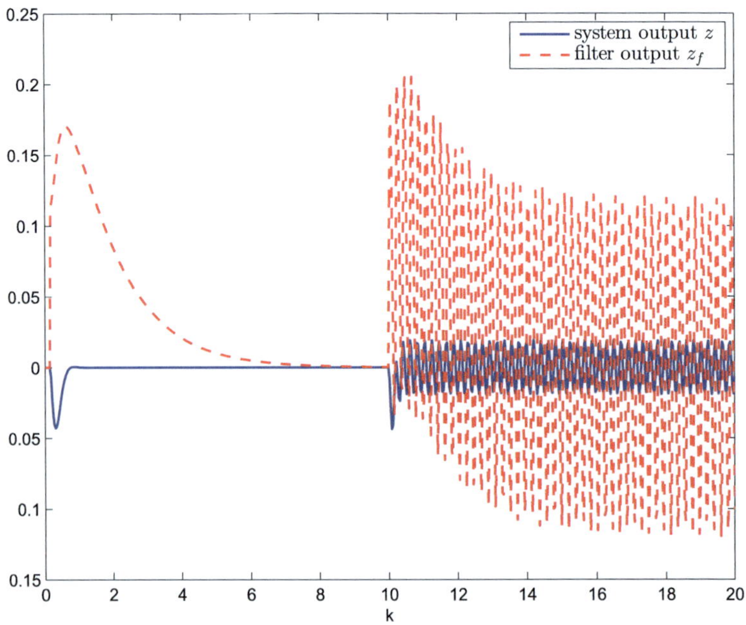

Fig. 3.4 System performance output z and filter performance output z_f

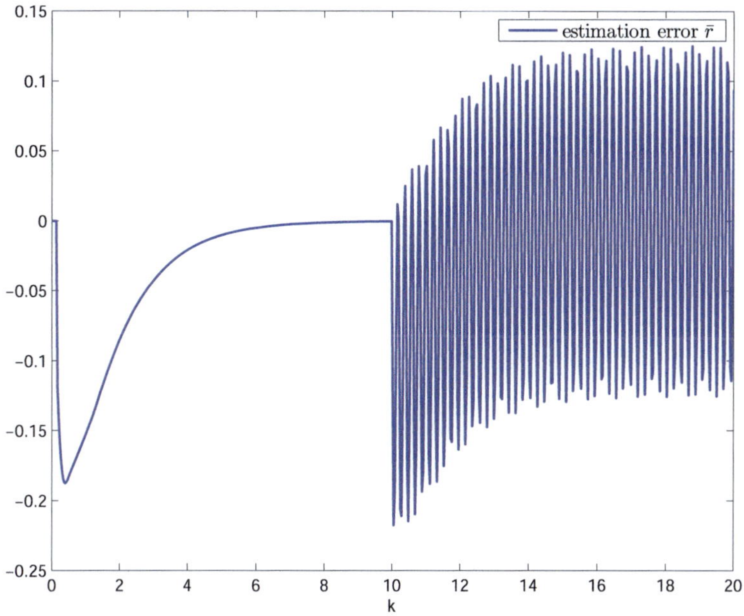

Fig. 3.5 Estimation error $\bar{r}$

3.5 Summary

In this chapter, the problem of H_∞ filter design for discrete-time T-S fuzzy systems with multiple time-varying delays has been studied. An improved delays-dependent region partitioning method that takes advantage of all possible information about the relationship between all the delay related terms and the current system states is proposed. Based on this method, a novel Lyapunov functional is introduced to establish the desired performance conditions of the filtering error system. Compared with the existing methods that use only the information of the relationship between the partial delay related items and the current system states, more information about delay related items is used in the proposed method. Thus, the conservatism can be further reduced. Finally, an example is given to verify the feasibility of the theory.

3.6 Notes

The network delay is usually considered to be one of the main causes of performance degradation, oscillation and even instability in engineering systems [1–3]. Therefore, it is important and meaningful to explore a new method for dealing with the multiple time-varying delays problems in response to the shortcomings in the existing results to achieve better filtering results, which motivates this paper. In addition, it should be noted that the results presented in this chapter are mainly constructed based on the ones in Ref. [24].

References

1. Wu TB, Li FB, Yang CH, Gui WH (2018) Event-based fault detection filtering for complex networked jump systems. IEEE/ASME Trans Mechatron 23(2):497–505
2. Wu HN (2006) Delay-dependent stability analysis and stabilization for discrete-time fuzzy systems with state delay: a fuzzy Lyapunov-Krasovskii functional approach. IEEE Trans Syst Man Cybern, Part B (Cybern) 36(4):954–962
3. Cheng G, Baoku S (2008) Delay-dependent robust H_∞ control of convex polyhedral uncertain fuzzy systems. J Syst Eng Electron 19(6):1191–1198
4. Jiang XF, Han QL (2006) Delay-dependent robust stability for uncertain linear systems with interval time-varying delay. Automatica 42(6):1059–1065
5. Li FB, Du CL, Yang CH, Gui WH (2018) Passivity-based asynchronous sliding mode control for delayed singular Markovian jump systems. IEEE Trans Autom Control 63(8):2715–2721
6. Xu SY, Lam J (2005) Improved delay-dependent stability criteria for time-delay systems. IEEE Trans Autom Control 50(3):384–387
7. Sun C, Wang FL, He XQ (2017) Robust fault-tolerant control for fuzzy delay systems with unmeasurable premise variables via uncertain system approach. Int J Innov Comput, Inform Control 13:823–846
8. Trinh H, Aldeen M (1994) On the stability of linear systems with delayed perturbations. IEEE Trans Autom Control 39(9):1948–1951
9. Chen WH, Zheng WX (2010) Delay-independent minimum dwell time for exponential stability of uncertain switched delay systems. IEEE Trans Autom Control 55(10):2406–2413
10. Banos A, Barreiro A (2009) Delay-independent stability of reset systems. IEEE Trans Autom Control 54(2):341–346
11. Aleksandrov AY, Hu GD, Zhabko AP (2014) Delay-independent stability conditions for some classes of nonlinear systems. IEEE Trans Autom Control 59(8):2209–2214
12. Kwon OM, Park JH (2004) On improved delay-dependent robust control for uncertain time-delay systems. IEEE Trans Autom Control 49(11):1991–1995
13. Mahmoud MS, Ismail A (2005) New results on delay-dependent control of time-delay systems. IEEE Trans Autom Control 50(1):95–100
14. Zhu SQ, Zhang CH, Cheng ZL, Feng J (2007) Delay-dependent robust stability criteria for two classes of uncertain singular time-delay systems. IEEE Trans Autom Control 52(5):880–885
15. Fridman E, Shaked U (2012) An improved stabilization method for linear time-delay systems. IEEE Trans Autom Control 47(11):1931–1937
16. Park JH (2011) A new delay-dependent criterion for neutral systems with multiple delays. J Comput Appl Math 136(1–2):177–184
17. He Y, Wu M, She JH (2006) Delay-dependent stability criteria for linear systems with multiple time delays. IEE Proc-Control Theory Appl 153(4):447–452
18. Dong HL, Wang ZD, Lam J, Gao HJ (2012) Fuzzy-model-based robust fault detection with stochastic mixed time delays and successive packet dropouts. IEEE Trans Syst Man Cybern, Part B (Cybern) 42(2):365–376
19. Ko KS, Lee WI, Park PG, Sung DK (2018) Delays-dependent region partitioning approach for stability criterion of linear systems with multiple time-varying delays. Automatica 87:389–394
20. Seuret A, Gouaisbaut F (2013) Wirtinger-based integral inequality: application to time-delay systems. Automatica 49(9):2860–2866

21. Chen J, Lin C, Chen B, Wang QG (2017) Fuzzy-model-based admissibility analysis and output feedback control for nonlinear discrete-time systems with time-varying delay. Inform Sci 412–413:116–131
22. Park PG, Ko JW, Jeong CK (2011) Reciprocally convex approach to stability of systems with time-varying delays. Automatica 47(1):235–238
23. Gao HJ, Zhao Y, Lam J, Chen K (2009) H_∞ fuzzy filtering of nonlinear systems with intermittent measurements. IEEE Trans Fuzzy Syst 17(2):291–300
24. Wang XL, Yang GH (2019) H_∞ filtering for T-S fuzzy systems with multiple time-varying delays: An improved delays-dependent region partitioning method. Inform Sci 481:368–380

Chapter 4
Event-Triggered Filtering for T-S Fuzzy Systems With Bounded Disturbances

4.1 Introduction

In Chap.3 and this chapter, the H_∞ filtering problems of discrete-time T-S fuzzy systems were studied respectively, and the external disturbances in systems are assumed to be energy bounded. However, it should be noted that the above mentioned results are obtained based on the traditional sampling data, that is, the data is sampled with a fixed period, which is named the time-triggered mechanism [1–3]. Although this mechanism is easy to implement in practice, sampling with fixed period will lead to congestion of communication bandwidth and waste of network resources [4–6]. Especially for microprocessors with limited processing resources, it is urgent to establish a more effective communication mechanism to reduce the transmission of redundant data. Event-triggered mechanism is produced under this background, and attracted the wide attention of researchers [7–11]. The state estimation problems of wireless sensor networks with limited communication capabilities were studied in Refs. [8, 10]. In [9], a switching method was established to solve the trigger interval-time in the event-triggered mechanism, so as to ensure that the time-interval of the event-triggered has a positive lower bound. A new event-triggered mechanism was proposed in Ref. [11], in which the parameter in the event-triggered condition is time-varying, and the design of distributed H_∞ filters was studied under the proposed event-triggered condition. At the same time, the event-triggered strategies have also been applied to the T-S fuzzy systems. For examples, Pan and Yang [12], Li et al. [13], Wang and Yang [14], Wang and Zhao [15].

Up to now, the research on the event-triggered problems of T-S fuzzy systems has achieved fruitful results. However, it should be pointed out that, the existing results for T-S fuzzy systems are obtained based on the periodic sampling data, that is, the measurement outputs are sampled first and then entered into the event-triggered unit. The emergence of sampling steps has increased the cost burden and is unnecessary compared to the event-triggered unit [16–18]. In addition, in the existing results, the controllers/filters designed in the absence of event-triggered mechanisms will

X.-L. Wang et al., *Robust Filtering and Fault Detection for T-S Fuzzy Systems*,
https://doi.org/10.1007/978-981-96-5818-3_4

no longer be suitable for the case with event-triggered mechanisms, that is, the co-design of controller/filter and event-triggered mechanism is necessary among the existing research methods, which is time-consuming and complex. Therefore, it is very necessary to construct a new filter design scheme to overcome the problems in the existing results.

This chapter studies the problem of robust filtering for continuous-time T-S fuzzy systems with bounded external disturbances via premise-region-dependent event-triggered mechanisms. First, attention is focused on the analysis and design of piecewise fuzzy filters without considering the event-triggered mechanisms. Then, based on the established piecewise fuzzy filter, event-triggered piecewise fuzzy filter with time regularization is designed, such that the new filtering error system is Zeno-free and L_∞ stable, in which new event-triggered conditions that depend on the different regions where the premise variables of the T-S fuzzy system and the piecewise fuzzy filter are located are designed. Unlike the existing results, the designed observer gains in the absence of event-triggered mechanisms are also applicable to the case with event-triggered mechanisms, while the ones in the existing methods need to be redesigned. Finally, a simulation example is applied to verify the effectiveness of the proposed filtering scheme.

4.2 System Modeling and Problem Statement

4.2.1 System Model

A class of continuous-time T-S fuzzy systems with bounded external disturbances is considered, as shown below:

$$\begin{aligned}\dot{x}(t) &= \sum_{i=1}^{r} h_i(q(x))(A_i x(t) + D_i w(t)) \\ y(t) &= \sum_{i=1}^{r} h_i(q(x))C_i x(t) \\ z(t) &= \sum_{i=1}^{r} h_i(q(x))E_i x(t)\end{aligned} \tag{4.1}$$

where $i \in \mathscr{E} \triangleq \{1, 2, \ldots, r\}$, and $q(x(t)) = \begin{bmatrix} q_1(x(t)) \; q_2(x(t)) \; \cdots \; q_p(x(t)) \end{bmatrix}$ is defined as the premise variables vector. $x(t) \in R^n$, $w(t) \in R^w$, $y(t) \in R^m$ and $z(t) \in R^z$ denote unmeasurable system states, bounded disturbances, measurement outputs and output signals to be estimated, respectively. A_i, D_i, C_i and E_i are system parameters. For the sake of brevity, x in membership functions instead of $x(t)$ is used later.

Define $h_i(q(x))$ represents the membership function, which is described below:

$$h_i(q(x)) = \frac{\prod_{j=1}^{r} \varsigma_{ij}(q_j(x))}{\sum_{j=1}^{r}\prod_{j=1}^{r} \varsigma_{ij}(q_j(x))},\ \eta_i(q(x)) = \prod_{j=1}^{r} \varsigma_{ij}(q_j(x)) \tag{4.2}$$

The premise variables of the system may not be available for the fuzzy filter implementation under the event-triggered communication scheme as some measured information involved premise variables may be not precisely received by the fuzzy filter under network environment. Therefore, the premise variables in T-S fuzzy system and the ones in fuzzy filter may not be synchronized. In order to solve this problem, the partition method is adopted in this chapter to deal with the asynchronous problem of the premise variables. Define Υ_l as the polyhedral region, which satisfies $\Upsilon_l \cap \Upsilon_s = \partial \Upsilon_{\varsigma}^{\iota}, l \neq s,\ l, s \in \ell,\ \varsigma \in \Re$, where $l \in \ell = \{1, 2, \ldots, \varrho\}$, and ℓ represents the polyhedral region indices. Then, the premise variable space $\mathscr{R}^s$ is decomposed into ϱ crisp regions and fuzzy regions according to the following decomposition criteria:

$$\partial \Upsilon_{\varsigma}^{\iota} = \left\{ q(x(t)) | h_{\varsigma}(x(t)) = 1, 0 \leq \underset{|\kappa| \longrightarrow 0^+}{h_{\varsigma}(x(t+\kappa))} < 1, \varsigma \in \mathscr{E} \right\} \tag{4.3}$$

where ι is the set of the face indices of the polyhedral hull and satisfies $\partial \Upsilon_{\varsigma} = \cup_{\iota}(\partial \Upsilon_{\varsigma}^{\iota})$.

Then, the global T-S fuzzy system (4.1) is equivalently modeled as a piecewise T-S fuzzy system, which is described below:

$$\begin{aligned} \dot{x}(t) &= A_l(h_t)x(t) + D_l(h_t)w(t) \\ y(t) &= C_l(h_t)x(t) \\ z(t) &= E_l(h_t)x(t),\ x(t) \in \Upsilon_l,\ l \in \ell \end{aligned} \tag{4.4}$$

where

$$A_l(h_t) = \sum_{m \in \ell(l)} h_m(q(x))A_m,\ D_l(h_t) = \sum_{m \in \ell(l)} h_m(q(x))D_m$$

$$C_l(h_t) = \sum_{m \in \ell(l)} h_m(q(x))C_m,\ E_l(h_t) = \sum_{m \in \ell(l)} h_m(q(x))E_m.$$

4.2.2 Piecewise Fuzzy Filter

The functions of the filter states are chosen as the PVs in the piecewise fuzzy filter because the system states are not measurable. Then, based on the model (4.1), the piecewise fuzzy filter is constructed, as shown below.

Filter Rule $\mathscr{E}^n$: If $q_1(\hat{x})$ is ϵ_{n1} and $q_2(\hat{x})$ is ϵ_{n2} and $\cdots$ and $q_p(\hat{x})$ is ϵ_{np}, then

$$\begin{aligned}
\dot{\hat{x}}(t) &= A_n \hat{x}(t) + L_s(\hat{y}(t) - y(t)) \\
\hat{y}(t) &= C_n \hat{x}(t) \\
\hat{z}(t) &= E_n \hat{x}(t)
\end{aligned} \tag{4.5}$$

where $q(\hat{x}) = \left[q_1(\hat{x})\ q_2(\hat{x}) \cdots q_p(\hat{x}) \right]$, $\hat{y}(t) \in R^m$ and $\hat{z}(t) \in R^z$ represent the premise variable vector, outputs, and performance outputs of the filter, respectively. Moreover, L_s are the gains.

Similarly, the fuzzy filter (4.5) can be equivalently modeled as a piecewise fuzzy filter, as follows:

$$\begin{aligned}
\dot{\hat{x}}(t) &= A_s(\hat{h}_t)\hat{x}(t) + L_s(\hat{y}(t) - y(t)) \\
\hat{y}(t) &= C_s(\hat{h}_t)\hat{x}(t) \\
\hat{z}(t) &= E_s(\hat{h}_t)\hat{x}(t),\ \hat{x}(t) \in \Upsilon_s,\ s \in \ell
\end{aligned} \tag{4.6}$$

where

$$A_s(\hat{h}_t) = \sum_{n \in \ell(s)} h_n(\hat{x}) A_n,\ C_s(\hat{h}_t) = \sum_{n \in \ell(s)} h_n(\hat{x}) C_n,\ E_s(\hat{h}_t) = \sum_{n \in \ell(s)} h_n(\hat{x}) E_n$$

Next, define $\tilde{x}(t) = x(t) - \hat{x}(t)$, $\xi(t) = \left[x^T(t)\ \tilde{x}^T(t) \right]^T$, and $\tilde{z}(t) = z(t) - \hat{z}(t)$, and then, the filtering error system is formed by (4.4) and (4.6), which is shown as follows:

$$\begin{cases}
\dot{\xi}(t) = A_{ls}\xi(t) + D_{ls}w(t) \\
\tilde{z}(t) = E_{ls}\xi(t)
\end{cases} \tag{4.7}$$

where

$$A_{ls} = \begin{bmatrix} A_l(h(t)) & 0 \\ A_l(h(t)) + L_s C_l(h(t)) - A_s(\hat{h}_t) - L_s C_s(\hat{h}_t) & A_s(\hat{h}_t) + L_s C_s(\hat{h}_t) \end{bmatrix},$$

$$D_{ls} = \begin{bmatrix} D_l(h(t)) \\ D_l(h(t)) \end{bmatrix}, E_{ls} = \left[E_l(h(t)) - E_s(\hat{h}_t)\ E_s(\hat{h}_t) \right].$$

Remark 4.1 The premise variables that depend on the system states in T-S fuzzy system (4.1) are not fully measurable, which is caused by technology and cost, etc. In this case, the continuous-time T-S fuzzy system (4.4) is partitioned based on the general state space rather than the measurable output space, and therefore, it cannot be reported that the piecewise T-S fuzzy system (4.4) and the piecewise fuzzy filter (4.6) always operate in the same region synchronously. The main motivation of the research on asynchronous filter design method is that the established piecewise fuzzy filter design scheme in the absence of event-triggered mechanisms is still valid when the introduction of event-triggered mechanism leads to the mismatch of the premise variables.

4.2.3 L_∞-Stability With Specified Performance

Problem: Given the piecewise T-S fuzzy system (4.4), the purpose is to design an asynchronous filter in the form of (4.6) with a guaranteed performance in the L_∞ sense. In other words, given a performance index $\beta > 0$, a piecewise fuzzy filter with structure (4.6) is established such that the filtering error system (4.7) is globally uniformly L_∞-stable with performance level β if (S.1) and (S.2) are satisfied:

(S.1) *Global uniform exponential stability*: The filtering error system (4.7) with $w(t) = 0$ is globally uniformly exponentially stable.

(S.2) *Output response for zero initial augmented states*: For zero initial states $\xi(t_0) = 0$, and bounded external disturbance $w(t)$, the following inequality

$$\|\tilde{z}(t)\| \leq \beta \|w(t)\|_\infty \tag{4.8}$$

holds for $t \geq t_0$.

Furthermore, the following two goals need to be achieved. (1) The mismatch of premise variables is overcome regardless of whether the event-triggered mechanism is introduced or not. (2) The external bounded disturbance is effectively suppressed, and the performance outputs are accurately estimated. To this end, two lemmas are first given to facilitate the establishment of the main results.

4.3 Piecewise Fuzzy Filter Analysis and Synthesis

Lemma 4.1 ([19]) *Consider a system with external disturbances $w(t)$ and performance outputs $z(t)$, which is shown below:*

$$\dot{\xi}(t) = M(t, \xi(t), w(t)) \tag{4.9}$$

$$z(t) = N(t, \xi(t)) \tag{4.10}$$

Assume that there is a differentiable function $V(t)$, and scalars β_1, β_2, β_3, β_4 and β_5, such that the following inequalities are satisfied for $t \geq 0$:

$$\beta_1 \|\xi(t)\|^2 \leq V(t, \xi(t)) \leq \beta_2 \|\xi(t)\|^2 \tag{4.11}$$

$$\mathscr{D}V(t, \xi(t)) M(t, \xi(t), w(t)) \leq -2\beta_3 (V(t, \xi(t)) - \beta_4 \|w(t)\|^2) \tag{4.12}$$

$$\|N(t, \xi(t))\|^2 \leq \beta_5 V(t, \xi(t)) \tag{4.13}$$

in where $\mathscr{D}V(t, \xi(t))$ *represents the derivative of* $V(t, \xi(t))$. *Then, the model (4.9) is globally uniformly* L_∞*-stable with performance level* $\sqrt{\beta_4\beta_5}$.

Lemma 4.2 (Projection lemma) ([20]) *For matrices* $Q = Q^T \in R^{n\times n}$, $K \in R^{k\times n}$, $S \in R^{m\times n}$, *the following linear matrix inequality*

$$Q + K^T X^T S + (K^T X^T S)^T < 0 \tag{4.14}$$

is solvable for the variable X if and only if

$$(1)\ K_\perp^T Q K_\perp < 0,\ \text{if } S_\perp = 0,\ K_\perp \neq 0 \tag{4.15}$$

$$(2)\ S_\perp^T Q S_\perp < 0,\ \text{if } K_\perp = 0,\ S_\perp \neq 0 \tag{4.16}$$

$$(3)\ K_\perp^T Q K_\perp < 0,\ S_\perp^T Q S_\perp < 0,\ \text{if } S_\perp \neq 0,\ K_\perp \neq 0 \tag{4.17}$$

in which $S_\perp$ *and* $K_\perp$ *are, respectively, the right null spaces of S and K.*

Next, based on the above analysis, design conditions of the piecewise fuzzy filter will be constructed, and the corresponding results are summarized as Theorem 4.1, as shown below.

Theorem 4.1 *For the system model (4.4), the filtering error system (4.7) satisfies (S.1) and (S.2), if there exist scalars* $\beta_1 > 0$, $\beta_2 \geq 0$, *matrices* $P > 0$, X_s, $G_{ls} = \begin{bmatrix} G_{ls}^{11} & 0 \\ 0 & G_s^{22} \end{bmatrix}$, *and* $V_{ls} = \begin{bmatrix} V_{ls}^{11} & 0 \\ 0 & G_s^{22} \end{bmatrix}$, *such that (4.18) and (4.19) hold with* $l, s \in \ell$:

$$\begin{bmatrix} -G_{ls} - G_{ls}^T & P + G_{ls}A_{ls} - V_{ls}^T & G_{ls}D_{ls} \\ \star & \beta_1 P + V_{ls}A_{ls} + A_{ls}^T V_{ls}^T & V_{ls}D_{ls} \\ \star & \star & -\beta_2 I \end{bmatrix} < 0 \tag{4.18}$$

$$\begin{bmatrix} P & E_{ls}^T \\ \star & -\beta_2 I \end{bmatrix} < 0 \tag{4.19}$$

Then, the gains of the piecewise fuzzy filter are calculated below:

$$L_s = (G_s^{22})^{-1} X_s$$

Proof The considered system model is a continuous-time T-S fuzzy model, and the partition approach is adopted to deal with the mismatch of premise variables. Therefore, it is necessary to consider the continuous crossing of regional boundaries. However, the affine terms are not contained in the considered model, thus, a constant quadratic Lyapunov function is introduced and established as follows:

$$V(t, \xi(t)) = \xi^T(t) P \xi(t) \tag{4.20}$$

Note that the Lyapunov function established above satisfies Rayleigh inequality, that is

$$\lambda_{min}(P)\|\xi(t)\|^2 \leq V(t, \xi(t)) \leq \lambda_{max}(P)\|\xi(t)\|^2$$

for any $\xi(t)$.

The time derivative of $V(t, \xi(t))$ calculated along the trajectory of the filtering error system (4.7) is

$$\mathscr{D}V(t, \xi(t))\dot{\xi}(t) = 2\xi^T(t)P(A_{ls}\xi(t) + D_{ls}w(t)) \tag{4.21}$$

Then, it follows from (4.21) that

$$\begin{aligned} &\mathscr{D}V(t, \xi(t))\dot{\xi}(t) + \beta_1 V(t, \xi(t)) - \beta_2\|w(t)\| \\ =&\begin{bmatrix} \xi(t) \\ w(t) \end{bmatrix}^T \begin{bmatrix} PA_{ls} + A_{ls}^T P + \beta_1 P & PD_{ls} \\ \star & -\beta_2 I \end{bmatrix} \begin{bmatrix} \xi(t) \\ w(t) \end{bmatrix} \end{aligned} \tag{4.22}$$

Furthermore, based on (4.22), L_∞-stable is ensured through the following inequality:

$$\begin{bmatrix} PA_{ls} + A_{ls}^T P + \beta_1 P & PD_{ls} \\ \star & -\beta_2 I \end{bmatrix} < 0, l, s \in \ell. \tag{4.23}$$

It is noted that the matrix inequality (4.23) is non-convex, thus, it is expensive to be solved globally from a computational point of view. Therefore, the main purpose of the following is to transform the product of the system matrices and the Lyapunov matrix, and the Projection Lemma is used to realize decoupling. For this purpose, the matrix inequality (4.23) is converted to linear matrix inequality, and the detailed calculation steps are given below.

First, the inequality (4.23) is rewritten as:

$$\underbrace{\begin{bmatrix} A_{ls} & D_{ls} \\ I & 0 \\ 0 & I \end{bmatrix}^T}_{U_\perp^T} \underbrace{\begin{bmatrix} 0 & P & 0 \\ \star & \beta_1 P & 0 \\ \star & \star & -\beta_1 I \end{bmatrix}}_{\mathscr{W}} \underbrace{\begin{bmatrix} A_{ls} & D_{ls} \\ I & 0 \\ 0 & I \end{bmatrix}}_{U_\perp} < 0 \tag{4.24}$$

where $U_\perp$ and $\mathscr{W}$ have been defined above according to the Projection Lemma.

Explicit null space calculation yields:

$$\begin{bmatrix} -I & A_{ls} & D_{ls} \end{bmatrix} = U, \quad \begin{bmatrix} A_{ls} & D_{ls} \\ I & 0 \\ 0 & I \end{bmatrix} = U_\perp \tag{4.25}$$

and define the following equations:

$$\begin{bmatrix} G_{ls}^T & V_{ls}^T & 0 \end{bmatrix}^T = \mathscr{X}, \mathscr{V} = I, \mathscr{V}_\perp = 0 \tag{4.26}$$

in which

$$G_{ls} = \begin{bmatrix} G_{ls}^{11} & 0 \\ 0 & G_s^{22} \end{bmatrix}, V_{ls} = \begin{bmatrix} V_{ls}^{11} & 0 \\ 0 & G_s^{22} \end{bmatrix}$$

Then, the following relationship is obtained by applying Projection Lemma:

$$\begin{bmatrix} 0 & P & 0 \\ \star & \beta_1 P & 0 \\ \star & \star & -\beta_1 I \end{bmatrix} + U^T \mathscr{X} \mathscr{V} + \mathscr{V}^T \mathscr{X} U \tag{4.27}$$
$$= \begin{bmatrix} -G_{ls} - G_{ls}^T & P + G_{ls} A_{ls} - V_{ls}^T & G_{ls} D_{ls} \\ \star & \beta_1 P + V_{ls} A_{ls} + A_{ls}^T V_{ls}^T & V_{ls} D_{ls} \\ \star & \star & -\beta_1 I \end{bmatrix}$$
$$< 0$$

Furthermore, substituting the matrices G_{ls} and V_{ls} presented in (4.26) into (4.27), together with the definition of $X_s = G_s^{22} L_s$, the final inequality (4.18) is established.

Note that, the inequality (4.18) means that $-G_{ls} - G_{ls}^T < 0$ holds, which shows that matrix G_{ls} is nonsingular. Thus, the gains of the piecewise fuzzy filter in each local region can be calculated by equation $L_s = (G_s^{22})^{-1} X_s$.

In addition, since $\beta_2 > 0$, Schur complement is applied to (4.19), one has

$$P - \beta_2^{-1} E_{ls}^T E_{ls} \geq 0 \tag{4.28}$$

and then, the above inequality pre- and post multiplied by $\xi(t)$ and its transpose, to get

$$\xi^T(t) E_{ls}^T E_{ls} \xi(t) - \beta_2 \xi^T(t) P \xi(t) \leq 0 \tag{4.29}$$

which means that

$$\| E_{ls} \xi(t) \|^2 \leq \beta_2 V(t, \xi(t)) \tag{4.30}$$

Thus, the inequality (4.13) holds with $N(t, \xi(t)) = E_{ls} \xi(t)$. The proof is completed. □

Remark 4.2 The main purpose of using partition method to handle asynchronous problem of premise variables is to design event-triggered mechanism through filter gains solved by inequalities (4.18) and (4.19) in Theorem 4.1. Thus, the conclusion in Theorem 4.1 needs to be still valid when event-triggered mechanism is introduced.

4.4 Event-Triggered Filtering Design

In this section, the new event-triggered mechanisms that depend on the different regions where the premise variables of the T-S fuzzy system and the piecewise fuzzy filter are located will be designed based on the piecewise fuzzy filter established in the previous section. Due to the fact that the established piecewise fuzzy filter (4.6) in the absence of event-triggered mechanisms is also applicable to the case with event-triggered mechanisms, the inputs of the fuzzy filter update the data only at certain triggering instances $t_0\ t_1\ \cdots\ t_k < t_{k+1}$, and keeps the current outputs inconvenient until the next successive triggering instance. Define $y_s(t) = y(t_k)$, $\forall t \in [t_k, t_{k+1})$, and then, the piecewise fuzzy filter (4.6) is re-presented as follows:

$$\begin{aligned} \dot{\hat{x}}(t) &= A_s(\hat{h}_t)\hat{x}(t) + L_s(\hat{y}(t) - y_s(t)) \\ \hat{y}(t) &= C_s(\hat{h}_t)\hat{x}(t) \\ \hat{z}(t) &= E_s(\hat{h}_t)\hat{x}(t),\ \hat{x}(t) \in \Upsilon_s,\ s \in \ell \end{aligned} \tag{4.31}$$

in which the gains L_s has been designed in Theorem 4.1, and the filter is the same as that in (4.6), except that it is updated at the triggering instances, which is determined by the event-triggered rule in the following form:

$$t_{k+1} = inf\{t | t \geq t_k + \tau_{ls}, \|y_e(t)\| > \sigma_{ls}\|y(t)\| + \epsilon_{ls}\} \tag{4.32}$$

in which $y_e(t)$ is represented as $y_e(t) = y_s(t) - y(t)$. τ_{ls} is positive Number to specify, and $\sigma_{ls} > 0$ and ϵ_{ls} are thresholds, in which σ_{ls} will be defined later. The time update rule (4.32) ensures that the lower bound of internal execution time $t_{k+1} - t_k$ is a positive constant τ_{ls}, which means that the Zeno phenomenon will not occur.

Remark 4.3 Different from the existing results, the piecewise fuzzy filter is first designed in the absence of event-triggered mechanisms, and then the above-mentioned event-triggered rule is established based on the designed filter. Thus, the designed filter gains in the absence of event-triggered mechanisms are also applicable to the case with event-triggered mechanisms, while the ones in the existing methods need to be re-designed. In addition, the proposed event-triggered mechanism is different from the existing results because parameters τ_{ls}, σ_{ls} and ϵ_{ls} depend on the different regions where the premise variables of the T-S fuzzy system and the piecewise fuzzy filter are located.

The new filtering error system established by combining system model (4.1) and event-triggered piecewise fuzzy filter (4.31) is described as follows:

$$\begin{cases} \dot{\xi}(t) = A_{ls}\xi(t) + D_{ls}w(t) + H_{ls}y_e(t) \\ \tilde{z}(t) = E_{ls}\xi(t) \end{cases} \tag{4.33}$$

in which $H_{ls} = \begin{bmatrix} 0 & L_s^T \end{bmatrix}^T$, and other parameters have been defined in (4.7).

Next, based on the above analysis, the main results will be given and summarized as Theorem 4.2.

Theorem 4.2 *Consider the system model (4.1) with $\|w(t)\|_\infty \leq w_0$, where w_0 is a positive constant. Assume that there are scalars $\beta_1 > 0$, $\beta_2 \geq 0$, and matrices $P > 0$, gain matrices L_s, such that the filtering error system (4.33) satisfies $\dot{V}(t,\xi(t)) \leq -\beta_1 V(t,\xi(t)) + \beta_2\|w(t)\|^2$, in which $V(t,\xi(t)) = \xi^T(t)P\xi(t)$. Select $\epsilon_{ls} > 0$, and*

$$\sigma_{ls} = \frac{\rho\beta_1\lambda_{min}(P)}{2\|PH_{ls}\|\|C\|} \tag{4.34}$$

in which $0 < \rho < 1$. Define $\tau_{ls} > 0$ as the solution of equation $\phi(\tau_{ls}) = 1$, where ϕ is the solution of the following ordinary differential equation:

$$\begin{aligned}\dot{\phi} &= \left(\frac{\|A_{ls}\|}{\sigma_{ls}} + \frac{\|C_{ls}\|\|D_{ls}\|w_0}{\epsilon_{ls}} + \|C_{ls}\|\|H_{ls}\|\phi\right)(1+\sigma_{ls}\phi),\\ \phi(0) &= 0\end{aligned} \tag{4.35}$$

Then, the filtering error system (4.33) implementing the event-triggered rule (4.32) is L_∞-stable.

Proof Since the derivative of $V(t,\xi(t))$ along the trajectory of the filtering error system (4.7) satisfies $\dot{V}(t,\xi(t)) \leq -\beta_1 V(t,\xi(t)) + \beta_2\|w(t)\|^2$, then the derivative of $V(t,\xi(t))$ along the filtering error system (4.33) satisfies:

$$\dot{V}(t,\xi(t)) \leq -\beta_1 V(t,\xi(t)) + \beta_2\|w(t)\|^2 + 2\xi^T(t)PH_{ls}y_e(t) \tag{4.36}$$

Since $\|\xi(t)\| \geq \|x(t)\| \geq \frac{\|y(t)\|}{\|C_{ls}\|}$, then the following inequality:

$$\|y_e(t)\| < \sigma_{ls}\|y(t)\| + \epsilon_{ls} \tag{4.37}$$

is rewritten as:

$$\|y_e(t)\| \leq \sigma_{ls}\|C_{ls}\|\|\xi(t)\| + \epsilon_{ls} \tag{4.38}$$

In this way, one has

$$\begin{aligned}\dot{V}(t,\xi(t)) \leq& -\beta_1 V(t,\xi(t)) + \beta_2\|w(t)\|^2 + 2\|\xi^T(t)\|\|PH_{ls}\|\sigma_{ls}\|C_{ls}\|\|\xi(t)\| \\ &+ 2\|\xi^T(t)\|\|PH_{ls}\|\epsilon_{ls}\end{aligned} \tag{4.39}$$

Next, according to (4.34), the following inequality is further constructed:

$$\dot{V}(t,\xi(t)) \leq -((1-\rho)\beta_1\lambda_{min}(P) - c_1)\|\xi(t)\| + \beta_2\|w(t)\|^2 + \frac{(\|PH_{ls}\|\epsilon_{ls})^2}{c_1} \tag{4.40}$$

in which $0 < c_1 < (1-\rho)\beta_1\lambda_{min}(P)$.

Furthermore, from (4.33) one obtains:

$$\|\dot{\xi}(t)\| \leq \|A_{ls}\|\|\xi(t)\| + \|D_{ls}\|\|w(t)\| + \|H_{ls}\|\|y_e(t)\| \tag{4.41}$$

and

$$\|\dot{y}_e(t)\| \leq \|\dot{y}(t)\| \leq \|C_{ls}\|\|\dot{x}(t)\| \leq \|C_{ls}\|\|\dot{\xi}(t)\| \tag{4.42}$$

Next, let $m(t) = \frac{\|y_e(t)\|}{\sigma_{ls}\|C_{ls}\|\|\xi(t)\|+\epsilon_{ls}}$, and then, for any $h > 0$, one has

$$\begin{aligned} m(t+h) - m(t) =& \frac{\|y_e(t+h)\|}{\sigma_{ls}\|C_{ls}\|\|\xi(t+h)\| + \epsilon_{ls}} - \frac{\|y_e(t)\|}{\sigma_{ls}\|C_{ls}\|\|\xi(t)\| + \epsilon_{ls}} \\ =& \frac{(\|y_e(t+h)\| - \|y_e(t)\|)(\sigma_{ls}\|C_{ls}\|\|\xi(t)\| + \epsilon_{ls})}{(\sigma_{ls}\|C_{ls}\|\|\xi(t+h)\| + \epsilon_{ls})(\sigma_{ls}\|C_{ls}\|\|\xi(t)\| + \epsilon_{ls})} \\ &- \frac{\sigma_{ls}\|C_{ls}\|\|y_e(t)\|(\|\xi(t+h)\| - \|\xi(t)\|)}{(\sigma_{ls}\|C_{ls}\|\|\xi(t+h)\| + \epsilon_{ls})(\sigma_{ls}\|C_{ls}\|\|\xi(t)\| + \epsilon_{ls})} \end{aligned} \tag{4.43}$$

and thus, the following equation is constructed:

$$\begin{aligned} D^+m(t) =& \lim_{h\to 0^+} \frac{m(t+h) - m(t)}{h} \\ =& \frac{D^+y_e(t)}{\sigma_{ls}\|C_{ls}\|\|\xi(t+h)\| + \epsilon_{ls}} - \frac{\sigma_{ls}\|C_{ls}\|\|y_e(t)\|D^+\|\xi(t)\|}{(\sigma_{ls}\|C_{ls}\|\|\xi(t+h)\| + \epsilon_{ls})^2} \end{aligned} \tag{4.44}$$

Note that

(1) when $\xi(t) \neq 0$, $D^+\|\xi(t)\| = \frac{\xi^T(t)\dot{\xi}(t)}{\|\xi(t)\|}$, and therefore, $|D^+\|\xi(t)\|| < \|\dot{\xi}(t)\|$.

(2) when $\xi(t) = 0$, $D^+\|\xi(t)\| = \lim sup_{h\to 0^+} \frac{\xi(t+h)-\xi(t)}{h} = \lim sup_{h\to 0^+} \frac{\xi(t+h)}{h} = \|\dot{\xi}(t)\|$.

Thus, it can be summarized as follows: $|D^+\|\xi(t)\|| < \|\dot{\xi}(t)\|$. Similarly, one has $|D^+\|y_e(t)\|| < \|\dot{y}_e(t)\|$.

Based on the above analysis, the equation (4.44) is further refined as follows:

$$\begin{aligned} D^+m(t) \leq& \frac{\|C_{ls}\|\|\dot{\xi}(t)\|}{\sigma_{ls}\|C_{ls}\|\|\xi(t)\| + \epsilon_{ls}} + \frac{\sigma_{ls}\|C_{ls}\|\|y_e(t)\|\|\dot{\xi}(t)\|}{(\sigma_{ls}\|C_{ls}\|\|\xi(t+h)\| + \epsilon_{ls})^2} \\ =& \frac{D^+y_e(t)}{\sigma_{ls}\|C_{ls}\|\|\xi(t+h)\| + \epsilon_{ls}} - \frac{\sigma_{ls}\|C_{ls}\|\|y_e(t)\|D^+\|\xi(t)\|}{(\sigma_{ls}\|C_{ls}\|\|\xi(t+h)\| + \epsilon_{ls})^2} \\ =& \frac{\|C_{ls}\|\|\dot{\xi}(t)\|}{\sigma_{ls}\|C_{ls}\|\|\xi(t)\| + \epsilon_{ls}}\left(1 + \frac{\sigma_{ls}\|y_e(t)\|}{\sigma_{ls}\|C_{ls}\|\|\xi(t)\| + \epsilon_{ls}}\right) \end{aligned} \tag{4.45}$$

in which some of the formulas in the third line of (4.45) have the following equation relationship:

$$\frac{\|C_{ls}\|\|\dot{\xi}(t)\|}{\sigma_{ls}\|C_{ls}\|\|\xi(t)\|+\epsilon_{ls}}=\frac{\|C_{ls}\|(\|A_{ls}\|\|\xi(t)\|+\|D_{ls}\|\|w(t)\|+\|H_{ls}\|\|y_e(t)\|)}{\sigma_{ls}\|C_{ls}\|\|\xi(t)\|+\epsilon_{ls}} \tag{4.46}$$

In addition, note that the first term on the right in (4.46) satisfies:

$$\frac{\|C_{ls}\|\|A_{ls}\|\|\xi(t)\|}{\sigma_{ls}\|C_{ls}\|\|\xi(t)\|+\epsilon_{ls}}\leq\frac{\|A_{ls}\|}{\sigma_{ls}} \tag{4.47}$$

the second term satisfies:

$$\frac{\|C_{ls}\|\|H_{ls}\|\|y_e(t)\|}{\sigma_{ls}\|C_{ls}\|\|\xi(t)\|+\epsilon_{ls}}=\|C_{ls}\|\|H_{ls}\|m(t) \tag{4.48}$$

and the third term satisfies:

$$\frac{\|C_{ls}\|\|D_{ls}\|\|w(t)\|}{\sigma_{ls}\|C_{ls}\|\|\xi(t)\|+\epsilon_{ls}}\leq\frac{\|C_{ls}\|\|D_{ls}\|w_0}{\epsilon_{ls}} \tag{4.49}$$

Thus, substituting (4.46), (4.47), (4.48) and (4.49) into (4.45) yields:

$$\mathscr{D}^+m(t)\leq\left(\frac{\|A_{ls}\|}{\sigma_{ls}}+\frac{\|C_{ls}\|\|D_{ls}\|w_0}{\epsilon_{ls}}+\|C_{ls}\|\|H_{ls}\|m(t)\right)(1+\sigma_{ls}m(t)) \tag{4.50}$$

Since $m(t_k)=0$, $m(t)\leq\phi(t-t_k)$ is obtained from the comparison lemma. Moreover, because the lower bound of time for evolution of m from 0 to 1 is τ_{ls}, Eq. (4.38) remains unchanged in $[t_k, t_k+\tau_{ls})$. For any $k\geq 0$, if $t_{k+1}=t_k+\tau_{ls}$, then (4.38) remains unchanged during the above interval $[t_k, t_{k+1})$. If $t_{k+1}>t_k+\tau_{ls}$, then (4.37) holds in the interval $[t_k+\tau_{ls}, t_{k+1})$, which implies that (4.38) is satisfied. Thus, (4.38) is always holds for $k\geq 0$. In addition, since the satisfaction of inequality (4.38) implies the existence of inequality (4.40), it can be summarized as follows: the filtering error system (4.33) implementing the event-triggered rule (4.32) is L_∞-stable. □

Remark 4.4 In the proof of Theorem 4.2, the equation for τ_{ls} is clearly shown. It should be noted that the proof of theorem 4.2 is still valid when $\tau_{ls}' \in (0, \tau_{ls}]$. In addition, in event-triggered rule (4.32), the parameter ϵ_{ls} can be arbitrarily selected, but there are two aspects to be considered when selecting ϵ_{ls}. The first one is that the system bound increases with the increase of ϵ_{ls}. The second one is that the equation of τ_{ls} depends on ϵ_{ls}. When ϵ_{ls} approaches 0, τ_{ls} decreases to 0. Therefore, the parameter ϵ_{ls} in triggering rule should be selected appropriately to balance execution time and system performance.

Remark 4.5 In this chapter, the event-triggered mechanisms dependent on the premise region are proposed for the first time, which overcomes two difficulties. The first one is to establish Eq. (4.34) to effectively avoid Zeno behavior, and the second one is to establish event-triggered mechanisms that depend on the premise regions, so as to save network resources more effectively.

4.5 Numeral Example

In this section, Theorem 4.2 is verified by the tunnel diode circuit in Ref. [21], in which the circuit diagram of tunnel diode circuit has been shown in Fig. 2.1 of Chap. 2, and the dynamic equation is established, as shown below:

$$i_D(t) = 0.002v_D(t) + 0.01v_D^3(t)$$

Define $x_1(t) = v_C(t)$ and $x_2(t) = i_L(t)$, and then, the state equation is constructed as follows:

$$\begin{aligned} C\dot{x}_1(t) &= -0.002x_1(t) - 0.01x_1^3(t) + x_2(t) \\ L\dot{x}_2(t) &= -x_1(t) - Rx_2(t) + v(t) \\ y(t) &= x_1(t) \end{aligned}$$

where $C = 20\,\text{mF}$, $L = 1000\,\text{mH}$ and $R = 10\,\Omega$.

Similar to Ref. [21], the state equation mentioned above is modeled as a fuzzy model with the same form as (4.1), in which the corresponding parameters are given below:

$$\begin{aligned} A_1 &= \begin{bmatrix} -0.1000 & 50.0000 \\ -1.0000 & -10.0000 \end{bmatrix}, A_2 = \begin{bmatrix} -4.6000 & 50.0000 \\ -1.0000 & -10.0000 \end{bmatrix}, \\ D_1 &= \begin{bmatrix} 0.0093 \\ 0.0181 \end{bmatrix}, D_2 = \begin{bmatrix} 0.0091 \\ 0.0181 \end{bmatrix}, C_1 = \begin{bmatrix} 1 & 0 \end{bmatrix}, C_2 = \begin{bmatrix} 1 & 0 \end{bmatrix}, \\ E_1 &= \begin{bmatrix} 1 & 0 \end{bmatrix}, E_2 = \begin{bmatrix} 1 & 0 \end{bmatrix}. \end{aligned}$$

The membership function curves are shown in Fig. 4.1, where the space of premise variables is decomposed into crisp region Υ_2 and fuzzy region Υ_1 according to the region decomposition criteria (4.3), which are shown as follows:

$$\begin{aligned} \Upsilon_1 &= \{x(t) |\ |x_1(t)| > 3\} \\ \Upsilon_2 &= \{x(t) |\ |x_1(t)| < 3\} \end{aligned}$$

In addition, it should be pointed out that the premise variables of the piecewise fuzzy filter are different from those of the system, the membership functions in system are selected as the ones in piecewise fuzzy filter, and the functions of the states in system are replaced by the ones in piecewise fuzzy filter as the premise variables.

The purpose is to design a piecewise fuzzy filter to overcome the difficulty induced by the mismatch of the premise variables, and then to establish the event-triggered rules based on the designed piecewise fuzzy filter, so as to avoid Zeno behavior and make the filtering error system achieve globally uniformly L_∞-stable. Based on this view, the results are discussed in three steps, as follows.

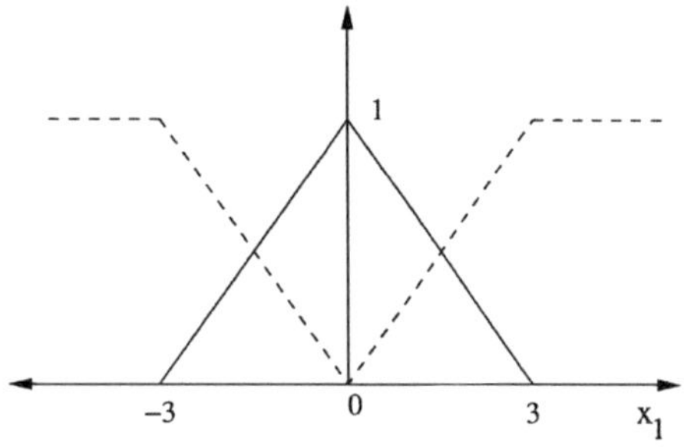

Fig. 4.1 Membership functions

Step 1: The corresponding parameters are selected as: $\beta_1 = 0.001$, $\beta_2 = 0.001$, $\rho = 0.5, \epsilon_{11} = 0.0001, \epsilon_{12} = 0.0001, \epsilon_{21} = 0.0001$ and $\epsilon_{22} = 0.0001$. Then, the performance index $\beta_2 = 1.3907$ is calculated, and the gains of the piecewise fuzzy filter and Lyapunov matrix are obtained by solving (4.18) and (4.19) as follows:

$$L_1 = \begin{bmatrix} -69.7040 \\ 12.2347 \end{bmatrix}, L_2 = \begin{bmatrix} -69.7040 \\ 12.2347 \end{bmatrix}$$

and

$$P = \begin{bmatrix} 0.3014 & -0.0421 & -0.0218 & -0.0047 \\ -0.0421 & 3.4844 & -0.0047 & 0.0113 \\ -0.0218 & -0.0047 & 1.1883 & -0.5316 \\ -0.0047 & 0.0113 & -0.5316 & 1.8210 \end{bmatrix}$$

Step 2: The event-triggered piecewise fuzzy filter is established in (4.31) with L_1 and L_2 solved above, then, let $w_0 = 2.4$, and thus, the parameters σ are, respectively, calculated as: $\sigma_{11} = 6.9948e - 07, \sigma_{12} = 6.9948e - 07, \sigma_{21} = 6.9948e - 07, \sigma_{22} = 6.9948e - 07$. In addition, in order to verify whether the designed event-triggered rule can avoid Zeno behavior, parameter τ can be obtained as $\tau_{11} = 8.0192e - 09$, $\tau_{12} = 7.6893e - 09$, $\tau_{21} = 8.0090e - 09$ and $\tau_{22} = 7.6982e - 09$ by solving the ordinary differential equation established in (4.35), in which note that $\tau_{11} > 0$, $\tau_{12} > 0$, $\tau_{21} > 0$ and $\tau_{22} > 0$, which indicates that there is Zeno-free under the proposed design method.

Step 3: Based on the above parameters, assume that the disturbance $w(t) = 2.4sin(0.2t)$ enters the T-S fuzzy system at $t = 0$, and then, the main results are given and shown in Figs. 4.2, 4.3, 4.4 and 4.5, in which Figs. 4.2 and 4.3 represent the state trajectories of the T-S fuzzy system and the piecewise fuzzy filter. The estimation error of performance output is plotted in Fig. 4.4. It can be seen from Fig. 4.4 that the performance output is accurately estimated under the designed scheme, and the error finally enters a small neighborhood of the origin as expected. Moreover, the inter-execution times $t_{k+1} - t_k$ of the event-triggered mechanism are presented in Fig. 4.5, the input of the piecewise fuzzy filter is shown in Fig. 4.6, and from Figs. 4.5 and

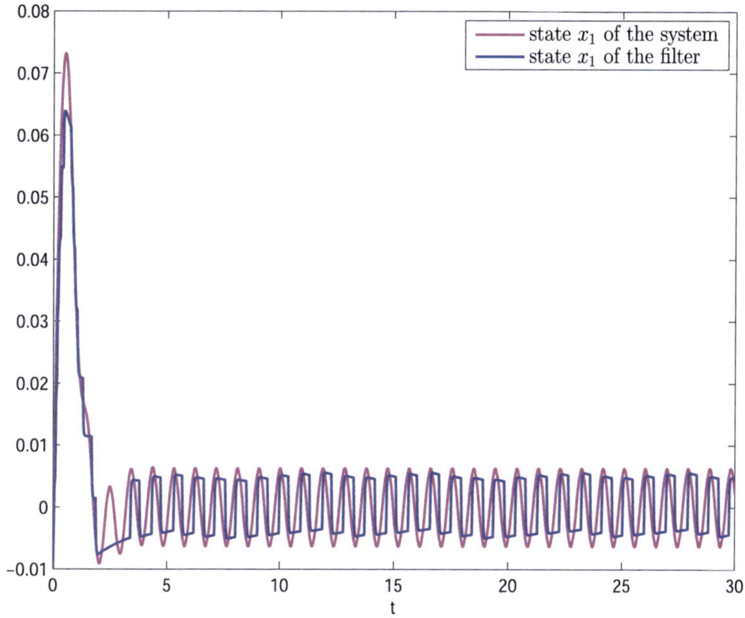

Fig. 4.2 System state x_1 and filter state x_1

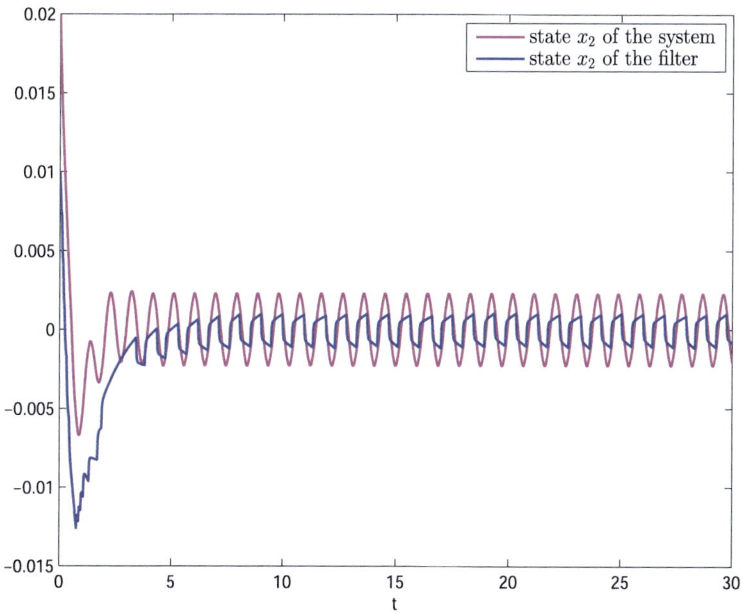

Fig. 4.3 System state x_2 and filter state x_2

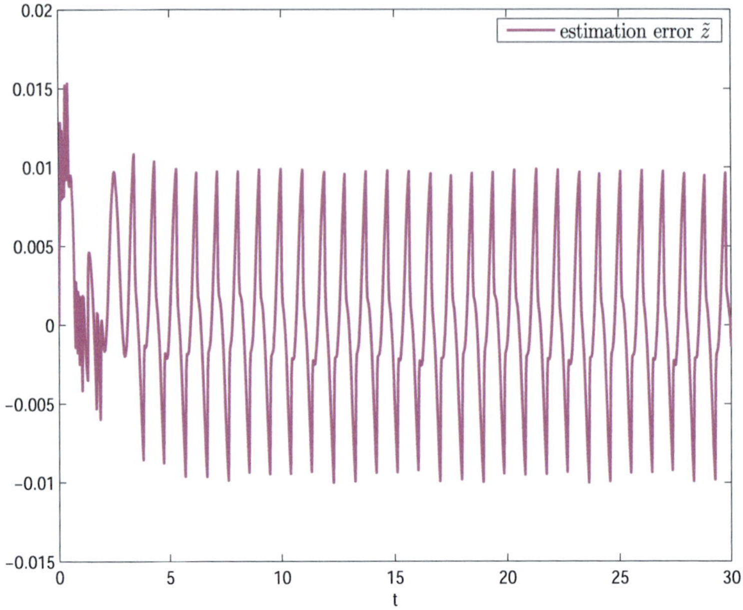

Fig. 4.4 Estimation errors $\tilde{z}$

Table 4.1 Minimum and average-execution time of event-triggered mechanism

Execution time	$\tau^{min}_{[0,20]}$	$\tau^{avg}_{[0,20]}$	$\tau^{min}_{[2,20]}$	$\tau^{avg}_{[2,20]}$
Event-triggered mechanism	0.030	0.4247	0.090	0.4473

4.6, it is indicated that the input $y_e(t)$ of the piecewise fuzzy filter is updated at each sampling instant t_k, which is determined by the established event-triggered rules.

In addition, assume $\tau^{min}_{[t^1,t^2]}$ and $\tau^{avg}_{[t^1,t^2]}$ as the minimum and average execution time within the interval $[t^1, t^2]$, respectively. Then, the parameters $\tau^{min}_{[0,20]}$ and $\tau^{avg}_{[2,20]}$ in the event-triggered rule (4.32) are calculated and displayed in Table 4.1. It can be seen from Table 4.1 that the minimum inter-execution time is 0.030 and the average-execution time in interval [2, 20] is 0.4473, which further shows the effectiveness of the designed event-triggered rule.

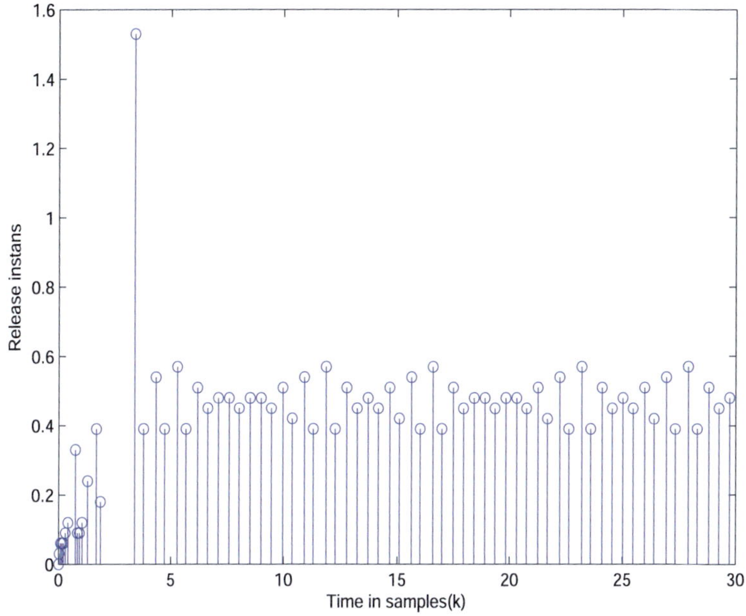

Fig. 4.5 Release instants and time intervals

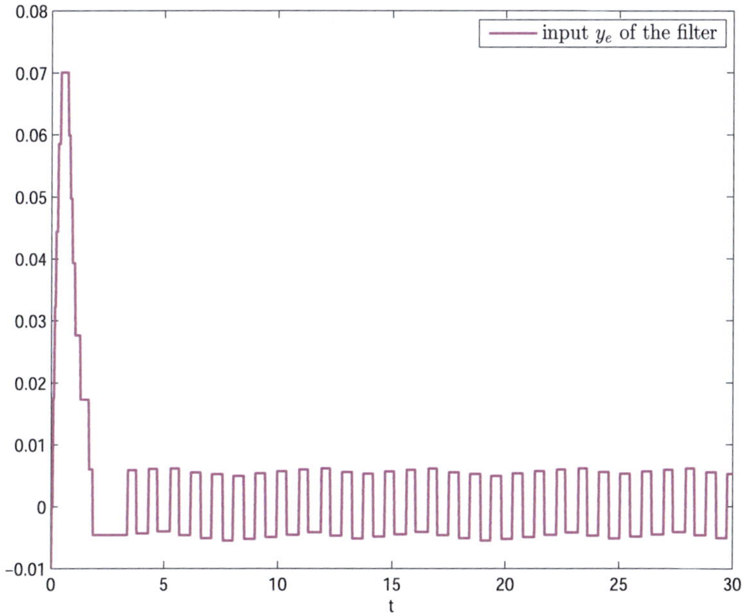

Fig. 4.6 Input of the filter

4.6 Summary

This chapter has investigated the robust filtering problem for continuous-time T-S fuzzy systems with bounded external disturbances via premise-region-dependent event-triggered mechanisms. First, piecewise fuzzy filter is constructed to overcome the difficulty induced by the mismatch of premise variables. Then, new event-triggered mechanisms that depend on the different regions where the premise variables of the T-S fuzzy system and the piecewise fuzzy filter are located are designed, and event-triggered piecewise fuzzy filter with time regularization is designed based on the proposed piecewise fuzzy filter, such that the new filtering error system is Zeno-free and L_∞ stable. Finally, a simulation example is applied to verify the effectiveness of the proposed filtering scheme.

4.7 Notes

The utilization rate of network resources can be effectively improved by introducing the event-triggered mechanisms, which will also result in the filters designed in the absence of event-triggered mechanisms will no longer be suitable for the case with event-triggered mechanisms. Therefore, it is very necessary to construct a new filtering scheme to solve the above-mentioned problems, which motivates this paper. In addition, it should be noted that the results presented in this chapter are mainly constructed based on the ones in Ref. [22].

References

1. Zhang J, Peng C, Xie XP, Yue D (2020) Output feedback stabilization of networked control systems under a stochastic scheduling protocol. IEEE Trans Cybern 50(6):2851–2860
2. Zhang XM, Han QL, Ge XH, Ding DR, Ding L, Yue D, Peng C (2020) Networked control systems: a survey of trends and techniques. IEEE/CAA J Automatica Sinica 7(1):1–17
3. Liu K, Fridman E, Hetel L (2012) Stability and L_2-gain analysis of networked control systems under round-robin scheduling: a time-delay approach. Syst Control Lett 61(5):666–675
4. Wang YL, Lim CC, Shi P (2017) Adaptively adjusted event-triggering mechanism on fault detection for networked control systems. IEEE Trans Cybern 47(8):2299–2311
5. Zhou WX, Fu J, Yan HC, Du X, Wang YY, Zhou H (2023) Event-triggered approximate optimal path-following control for unmanned surface vehicles with state constraints. IEEE Trans Neural Netw Learn Syst 34(1):104–118
6. Shi P, Wang HJ, Lim CC (2016) Network-based event-triggered control for singular systems with quantizations. IEEE Trans Indus Electron 63(2):1230–1238
7. Zong GD, Ren HL, Karimi HR (2021) Event-triggered communication and annular finite-time H_∞ filtering for networked switched systems. IEEE Trans Cybern 51(1):309–317
8. Ge XH, Han QL, Zhang XM, Ding L, Yang FW (2020) Distributed event-triggered estimation over sensor networks: a survey. IEEE Trans Cybern 50(3):1306–1320
9. Selivanov A, Fridman E (2016) Event-triggered H_∞ control: a switching approach. IEEE Trans Autom Control 61(810):3221–3226

10. Liu HB, Yu HS (2017) An event-triggered approach to robust state estimation for wireless sensor networks. J Control Decis 4(4):263–275
11. Ge XH, Han QL, Wang ZD (2019) A threshold-parameter-dependent approach to designing distributed event-triggered H_∞ consensus filters over sensor networks. IEEE Trans Cybern 49(4):1148–1159
12. Pan YN, Yang GH (2018) Event-triggered fault detection filter design for nonlinear networked systems. IEEE Trans Syst Man Cybern 48(11):1851–1862
13. Li HY, Chen ZR, Wu LG, Lam HK, Du HP (2017) Event-triggered fault detection of nonlinear networked systems. IEEE Trans Cybern 47(4):1041–1052
14. Wang XL, Yang GH (2018) Event-triggered fault detection for discrete-time T-S fuzzy systems. ISA Trans 76:18–30
15. Wang YZ, Zhao J (2022) Periodic event-triggered sliding mode control for switched uncertain T-S fuzzy systems with a logistic adaptive event-triggering scheme. IEEE Trans Fuzzy Syst 30(10):4115–4126
16. Li ZY, Zhao J (2022) A Zeno-free event-triggered control strategy for asymptotic stabilization of switched affine systems. IEEE Trans Autom Control 67(10):5509–5516
17. Dolk VS, Borgers DP, Heemels W (2017) Output-based and decentralized dynamic event-triggered control with guaranteed L_p-gain performance and Zeno-freeness. IEEE Trans Autom Control 66(1):34–49
18. Yu H, Chen Xia, Chen T W, Hao F (2020) Event-triggered bipartite consensus for multiagent systems: a Zeno-free analysis. IEEE Trans Autom Control 65(11):4866–4873
19. Chakrabarty A, Corless MJ, Buzzard GT, Zak SH, Rundell AE (2018) State and unknown input observers for nonlinear systems with bounded exogenous inputs. IEEE Trans Autom Control 62(11):5497–5510
20. Boyd S, El GL, Feron E, Balakrishnan V (1994) Linear matrix inequality in systems and control theory. SIAM, Philadelphia, PA
21. Gao HJ, Zhao Y, Lam J, Chen K (2009) H_∞ fuzzy filtering of nonlinear systems with intermittent measurements. IEEE Trans Fuzzy Syst 17(2):291–300
22. Wang XL, Yang GH (2022) Robust filtering for fuzzy systems with bounded disturbances via premise-region-dependent event-triggered mechanisms. IEEE Trans Fuzzy Syst 30(10):4093–4101

Chapter 5
Event-Triggered Fault Detection Observer Design for T-S Fuzzy Systems

5.1 Introduction

Modern control systems exhibit multiple characteristics such as high complexity, intelligence, and integration, etc., and a large number of configured actuators and sensors are prone to generate faults in complex and ever-changing working environments, resulting in more stringent requirements for the safety of the actual systems. Therefore, stricter requirements are put forward for the safety of the actual systems. Fault detection is very important to detect faults in time to improve the safety of actual systems, and has become a research hotspot [1–3], especially the research on event-triggered fault detection for the T-S fuzzy systems has been deeply carried out. For example, Refs. [4–8], in which in Ref. [4], an interval partitioning approach was proposed to relax the design conditions of the fault detection observers. In Ref. [5], the design conditions of the fault detection observers with less conservatism were also established. Moreover, in Ref. [6], a new fault detection scheme for the polynomial fuzzy systems was designed through the establishment of polynomial fault detection filter, in which the Taylor series is applied in the filter analysis to approximate the membership functions. In [7], H_∞ performance analysis conditions were derived by adopting the Lyapunov function analysis method with the slack matrices, and then the better fault detection performance is achieved.

Although the problems of event-triggered fault detection for T-S fuzzy systems have received widespread attention. It should be emphasized that the research on event-triggered fault detection for T-S fuzzy systems is facing a difficulty, that is, the mismatch of premise variables, which is caused by the introduction of event-triggered mechanism that makes the system information unable to be received by the fault detection observers in real time [9–11]. Thus, it is very important to study the asynchronous problems of the premise variables of T-S fuzzy systems in network environment, and several methods have been proposed through the efforts of researchers. The Refs. [5, 12] have dealt with the asynchronous problems of premise variables by using the partition method. However, the piecewise Lyapunov functions

X.-L. Wang et al., *Robust Filtering and Fault Detection for T-S Fuzzy Systems*,
https://doi.org/10.1007/978-981-96-5818-3_5

cannot be used in continuous-time T-S fuzzy systems within the technical framework of this method, which has the limitation of application. In Refs. [13–16], the non-PDC methods were designed to handle the mismatch of premise variables, in which there is a proportion relationship between the membership functions of the T-S fuzzy system and those of the observer. However, note that the constructed proportional relationship is an inequality one, which is uncertain, and the establishment of the membership functions of observer is not easy under the proposed technical framework. In Refs. [17, 18], the concept of matching membership function were proposed, and the stability and filtering of the systems were studied respectively. However, although in Refs. [17, 18], an equation involving the exchange information of membership function rules is established, the membership function rules are exchanged among different membership functions, which is obviously not tenable. Summarizing the above results, there are some limitations and defects in the existing methods. Therefore, in response to the shortcomings of the existing results in Refs. [13–18], it is very necessary to construct a novel strategy to deal with the mismatch of premise variables, thereby solving the design of event-triggered fault detection observers for T-S fuzzy systems.

This chapter investigates the design of event-triggered fault detection observers for T-S fuzzy systems. In order to provide design flexibility, the membership functions of the residual generator are different from those of the T-S fuzzy system and the fault detection observer. Unlike the existing asynchronous premise strategies that are designed based on inequality constraints between the membership functions in the systems and those in the residual generators, an improved matching membership function method is proposed, and the equality constraints are established under the proposed technical framework, such that the membership functions of the residual generator can be obtained directly without calculation. By applying the constraints of the fuzzy weighting parameters and the differences of each membership functions, new criteria in terms of linear matrix inequalities are derived to ensure the desired performance of the fault detection system. It is shown that the proposed method can overcome the shortcomings of the existing results. Finally, the effectiveness of the proposed fault detection scheme is verified by an example.

5.2 System Modeling and Problem Statement

A continuous-time T-S fuzzy model is considered, as shown below:

$$\begin{aligned} \dot{x}(t) &= \sum_{i=1}^{r} \mu_i(g(x))(A_i x(t) + D_i v(t) + F_i f(t)) \\ y(t) &= Cx(k) \end{aligned} \tag{5.1}$$

where $x(t) \in R^n$ and $y(t) \in R^m$ are, respectively, system states and measurable outputs. $v(t) \in R^v$ are external disturbances, and $f(t) \in R^f$ are actuator faults. r and

$g_q(x(t))$ $(q = 1, \ldots, p)$ represent the number of inference rules and the premise variables, respectively. A_i, D_i, F_i and C are system parameters. For the sake of brevity, x in membership functions instead of $x(t)$ is used later.

In addition, membership function is defined as $\mu_i(g(x))$ and described as follows:

$$\sum_{i=1}^{r} \mu_i(g(x)) = 1, \mu_i(g(x)) \geq 0, \forall i = 1, 2, \ldots, r. \tag{5.2}$$

The event-triggered mechanism is introduced to save network resources, and define $t_{q-1}h$ and $i_{q-1}h = t_{q-1}h + nh$ $(q = 1, 2, \ldots, N)$ as the current triggering instants and the sampling instants respectively, where h is the sampling period. The transmitted measurement outputs and the current outputs are, respectively, given as $y(t_{q-1}h)$ and $y(i_{q-1}h))$, and the next transmission instant is described as:

$$t_q h = t_{q-1}h + \min_{n \in N}\{nh | y_s^T \Omega y_s > \sigma y^T(t_{q-1}h)\Omega y(t_{q-1}h)\} \tag{5.3}$$

where $y_s = y(t_{q-1}h) - y(i_{q-1}h)$, $\Omega > 0$ and $\sigma \in [0,\ 1)$ denote the weight matrix and threshold respectively. The ZOHs are applied to keep the currently transmitted data unchanged until the next data arrival, and thus, we have $\bar{y}(t) = y(t_{q-1}h)$ for $t \in [t_{q-1}h, t_q h)$.

Next, define $[t_{q-1}h, t_q h) = \bigcup_{n=0}^{\varrho_{q-1}} \aleph_n$, where $\aleph_n = [t_{q-1}h + nh, t_{q-1}h + nh + h)$, $n = 0, 1, \ldots, \varrho_{q-1}$, and $\varrho_{q-1} = t_q - t_{q-1} - 1$. And define $\tau(t)$ and $e(t)$ as:

$$\tau(t) = \begin{cases} t - t_{q-1}h,\ t \in \aleph_0, \\ t - t_{q-1}h - \varrho_{q-1}h,\ t \in \aleph_{\varrho_{q-1}} \end{cases} \tag{5.4}$$

$$e(t) = \begin{cases} y(t_{q-1}h) - y(t_{q-1}h),\ t \in \aleph_0, \\ y(t_{q-1}h) - y(t_{q-1}h + \varrho_{q-1}h),\ t \in \aleph_{\varrho_{q-1}} \end{cases} \tag{5.5}$$

in which $0 \leq \tau(t) \leq h$. Then, one has

$$y(t_{q-1}h) = y(t - \tau(t)) + e(t) \tag{5.6}$$

Next, an event-triggered fuzzy fault detection observer is established to generate residual signals, in which the same membership functions as the T-S fuzzy system are selected as those of the fault detection observer, and the functions of the fault detection observer states are chosen as the premise variables of the fault detection observer. The specific results are presented below.

If $g_1(\hat{x})$ is ϵ_{j1} and $g_2(\hat{x})$ is ϵ_{j2} and $\cdots$ and $g_p(\hat{x})$ is ϵ_{jp}, then

$$\begin{aligned} \dot{\hat{x}}(t) &= A_j \hat{x}(t) + L_j(y(t_{q-1}h) - \hat{y}(t)) \\ \hat{y}(t) &= C\hat{x}(t),\ \ t \in [t_{q-1}h,\ t_q h) \end{aligned} \tag{5.7}$$

where L_j, $j = 1, 2, \ldots, r$ are the gains of the fault detection observer.

Then, the global event-triggered fuzzy fault detection observer is established, as shown below:

$$\begin{aligned}\dot{\hat{x}}(t) &= \sum_{j=1}^{r} \mu_j(g(\hat{x}))(A_j\hat{x}(t) + L_j(y(t_{q-1}h) - \hat{y}(t))) \\ \hat{y}(t) &= C\hat{x}(t), \quad t \in [t_{q-1}h,\ t_q h)\end{aligned} \tag{5.8}$$

It is known that the introduction of event-triggered mechanism will lead to the asynchronous problem of the premise variables. In this case, although the non-PDC method proposed in Refs. [14, 16] can deal with the mismatch of premise variables, there is an inequality relationship between the membership functions in T-S fuzzy system and the ones in observer, and the membership functions of the observer are not easily constructed within the constructed technical framework. In view of the above problems, a new method will be proposed, and the equality constraints will also be established. First, the functions of the observer states are chosen as the premise variables of the residual generator. Then, the established membership functions in residual generator and the ones in T-S fuzzy system are affine matched. The details are given below.

If $g_1(\hat{x})$ is ϵ_{l1} and $g_2(\hat{x})$ is ϵ_{l2} and $\cdots$ and $g_p(\hat{x})$ is ϵ_{lp}, then

$$r(t) = W_l(\bar{y}(t) - \hat{y}(t)), \quad t \in [t_{q-1}h,\ t_q h) \tag{5.9}$$

where W_l, $l = 1, 2, \ldots, r$ are the gains of the residual generator.

Then, the residual generator is established, as shown below:

$$r(t) = \sum_{l=1}^{r} h_l(g(\hat{x}))W_l(\bar{y}(t) - \hat{y}(t)), \quad t \in [t_{q-1}h,\ t_q h) \tag{5.10}$$

The established membership functions of the residual generator and those of the T-S fuzzy system are affine matched. Thus, the relationship between the membership functions in fault detection observer and those in residual generator is established, as shown below:

$$h_l(g(\hat{x})) = \alpha\mu_l(g(\hat{x})) + \beta,\ l = 1, 2, \ldots, r \tag{5.11}$$

in which α and β are scale coefficient and offset coefficient respectively.

Remark 5.1 In the existing results in Refs. [14, 16], the membership functions $\mu_j(\hat{x})$ in the controller do not match the ones $h_j(\hat{x})$ in the T-S fuzzy system, and the inequality constraint $h_j(\hat{x}) - \alpha\mu_j(\hat{x}) > 0$ is established by introducing a slack matrix. It is noted that this inequality constraint is uncertain, and the establishment of membership functions $h_j(\hat{x})$ in the controller is not easy. Unlike the existing results, an equality constraint is constructed by proposing the matching membership function method, which is a deterministic constraint, and the membership functions $h_j(\hat{x})$ are easy to be established.

Remark 5.2 In Ref. [18], the concept of matching membership function is proposed, and the stability criteria of the T-S fuzzy system are established. However, the exchange of membership function rules cannot be implemented under the technical framework constructed in Ref. [18]. Moreover, in Ref. [17], an equation involving the exchange information of membership function rules is established. However, the membership function rules are exchanged among different membership functions, which is obviously not tenable.

Synthesizing equations (5.1), (5.8) and (5.10), the following fault detection system is established:

$$\begin{aligned}\dot{\xi}(t) &= \sum_{i=1}^{r}\sum_{j=1}^{r}\mu_i(x)\mu_j(\hat{x})(\bar{A}_{ij}^q\xi(t)+\bar{B}_{ij}^q\xi(k-\tau(t))+\bar{E}_{ij}^q e(t)+\bar{D}_{ij}^q w(t))\\ \bar{r}(t) &= \sum_{l=1}^{r}h_l(\hat{x})(\bar{C}_l^q\xi(t)+\bar{G}_l^q\xi(k-\tau(t))+W_l^q e(t)+\bar{H}w(t))\end{aligned} \tag{5.12}$$

where $\xi(t)=\left[x^T(t)\ \tilde{x}^T(t)\right]^T$, $\tilde{x}(t)=x(t)-\hat{x}(t)$, $w(t)=\left[v^T(t)\ f^T(t)\right]^T$, $\bar{r}(t)=r(t)-Mf(t)M=\left[1\ \cdots\ 1\right]^T$, and the corresponding parameters are given below:

$$\begin{aligned}\bar{A}_{ij} &= \begin{bmatrix} A_i & 0\\ A_i-A_j+L_jC & A_j-L_jC\end{bmatrix}, \bar{E}_{ij}=\begin{bmatrix}0\\ -L_j\end{bmatrix}, \bar{B}_{ij}=\begin{bmatrix}0 & 0\\ -L_jC & 0\end{bmatrix},\\ \bar{C}_l &= \begin{bmatrix}-W_lC\ W_lC\end{bmatrix}, \bar{H}=\begin{bmatrix}0\ -M\end{bmatrix}, \bar{G}_l=\begin{bmatrix}W_lC\ 0\end{bmatrix}, \bar{D}_{ij}=\begin{bmatrix}D_i & F_i\\ D_i & F_i\end{bmatrix}\end{aligned}$$

The purpose is to establish the fault detection observer and residual generator for T-S fuzzy system (5.1) to detect the faults, and this purpose can be transformed into dealing with the following problems:

Problem: For the T-S fuzzy system (5.1), the fuzzy fault detection observer (5.8) and the residual generator (5.10) are constructed such that (Q.1) and (Q.2) are satisfied:

(Q.1) *Asymptotic Stability*: the fault detection system (5.12) with $w(t)=0$ is asymptotically stable.

(Q.2) H_∞ Performance: the fault detection system (5.12) with $w(t)\neq 0$ satisfies the following inequality:

$$\|\bar{r}(t)\|_2 \leq \lambda\|w(t)\|_2 \tag{5.13}$$

where λ is the H_∞ performance index.

In addition, the event triggered fault detection observer designed for the T-S fuzzy system needs to achieve the following two goals: (1) The asynchronous problem between the premise variables in T-S fuzzy system and the ones in fault detection observer is solved. (2) The faults can be successfully detected by the designed event-triggered fault detection observer.

5.3 Event-Triggered Fault Detection Observer Design

In this section, by defining an appropriate Lyapunov function, H_∞ performance analysis results are first obtained by applying the decision variable separation technique with free weighting matrices. Then, the gains of the fault detection observer and the residual generator are derived by the design conditions of the fault detection observer.

5.3.1 H_∞ Fault Detection Performance Analysis Conditions

Next, H_∞ performance analysis results are presented and summarized as Theorem 5.1, as follows.

Theorem 5.1 *For the given parameters $h > 0$, $0 \le \sigma < 1$, $0 \le \alpha \le 1$ and $\beta = \frac{1-\alpha}{r}$, if there exist variables $P > 0$, $Q > 0$, $N > 0$, $\Lambda_a > 0$, $\Lambda_b > 0$, $\Lambda_i > 0$, $i = 1, 2, \ldots, r$, Z, K, and $\lambda > 0$, the following inequalities are established for $i = 1, 2, \ldots, r$:*

$$\Pi_{ijl} - K < 0 \tag{5.14}$$

$$\begin{bmatrix} N & Z \\ Z^T & N \end{bmatrix} \ge 0 \tag{5.15}$$

$$X_i = \begin{bmatrix} X_{i11} & X_{i12} & \cdots & X_{i1r} & \Gamma_{i1} \\ \star & X_{i22} & \cdots & X_{i2r} & \Gamma_{i2} \\ \star & \star & \ddots & \vdots & \vdots \\ \star & \star & \star & X_{irr} & \Gamma_{ir} \\ \star & \star & \star & \star & \Upsilon_i \end{bmatrix} < 0 \tag{5.16}$$

where

$$X_{ijl} = \begin{cases} \frac{1}{2}\alpha(\Pi_{ijj} + \Pi_{ijj}) + o_{ijj}, & j = l = 1, 2, \ldots, r \\ \frac{1}{2}\alpha(\Pi_{ijl} + \Pi_{ilj}) + o_{ijl}, & j \ne l = 1, 2, \ldots, r \end{cases}$$

$$\Delta_{ijl} = \begin{cases} -\Lambda_a - 2\Lambda_j - j\Lambda_b, & j = l < r \\ -\Lambda_a + \Lambda_b, & j \ne l = r \\ \frac{1}{2}\alpha(\Pi_{ijl} + \Pi_{ilj}) + K - \alpha K + \Delta_{ijl}, & j \ne l \end{cases}$$

$$\Gamma_{ij} = \Lambda_a + \Lambda_j, \Upsilon_i = -\Lambda_a + \frac{r(r-1)}{2}\alpha\Lambda_b,$$

$$o_{ijj} = K - \alpha K + \Delta_{ijj}, o_{ijl} = K - \alpha K + \Delta_{ijl}$$

in which Π_{ijl} is described as:

$$\Pi_{ijl} = \begin{bmatrix} \Pi_{ijl}^1 & \Pi_{ijl}^2 \\ \star & \Pi_{ijl}^4 \end{bmatrix}$$

and

$$\Pi_{ijl}^1 = \begin{bmatrix} \bar{A}_{ij}^T P + P\bar{A}_{ij} + Q - N & P\bar{B}_{ij} + N + Z & -Z & P\bar{E}_{ij} \\ \star & -2N - Z^T - Z & Z + N & 0 \\ \star & \star & -Q - N & 0 \\ \star & \star & \star & -\Omega \end{bmatrix},$$

$$\Pi_{ijl}^2 = \begin{bmatrix} P\bar{D}_{ij} & h\bar{A}_{ij}^T P & \bar{C}_l^T & 0 \\ 0 & h\bar{B}_{ij}^T P & \bar{G}_l^T & \sqrt{\sigma}(CH_1)^T\Omega \\ 0 & 0 & 0 & 0 \\ 0 & h\bar{E}_{ij}^T P & \bar{W}_l^T & \sqrt{\sigma}\Omega \end{bmatrix},$$

$$\Pi_{ijl}^4 = \begin{bmatrix} -\lambda^2 I & h\bar{D}_{ij}^T P & \bar{H}^T & 0 \\ \star & N - P - P^T & 0 & 0 \\ \star & \star & -I & 0 \\ \star & \star & \star & -\Omega \end{bmatrix},$$

$$H_1 = \begin{bmatrix} I & 0 \end{bmatrix}.$$

Then, the fault detection system (5.12) satisfies the performances (Q.1) and (Q.2).

Proof For the fault detection system (5.12), the following Lyapunov function is established in interval $t \in [t_{q-1}h, t_q h)$:

$$V(t) = \xi^T(t)P\xi(t) + \int_{t-h}^{t} \xi^T(s)Q\xi(s)ds + h\int_{-h}^{0}\int_{t+\theta}^{t} \dot{\xi}^T(s)N\dot{\xi}(s)dsd\theta \tag{5.17}$$

where $P > 0$, $Q > 0$ and $N > 0$ are the parameters to be solved.

For (5.17), the time derivative of $V(t)$ yields:

$$\begin{aligned}\dot{V}(t) &= \dot{\xi}^T(t)P\xi(t) + \xi^T(t)P\dot{\xi}(t) + \xi^T(t)Q\xi(t) - \xi^T(t-h)Q\xi(t-h) \\ &\quad + h^2\dot{\xi}^T(t)N\dot{\xi}(t) - h\int_{t-h}^{t} \dot{\xi}^T(s)N\dot{\xi}(s)ds\end{aligned} \tag{5.18}$$

Through inequality (5.14) in Theorem 5.1, the following inequality relationship is established:

$$
\begin{aligned}
&-h\int_{t-h}^{t}\dot{\xi}^T(s)N\dot{\xi}(s)ds\\
&\leq -\begin{bmatrix}\xi(t-\tau(t))-\xi(t)\\ \xi(t-\tau(t))-\xi(t-h)\end{bmatrix}^T\begin{bmatrix}N & Z\\ Z^T & N\end{bmatrix}\begin{bmatrix}\xi(t-\tau(t))-\xi(t)\\ \xi(t-\tau(t))-\xi(t-h)\end{bmatrix}
\end{aligned}\tag{5.19}
$$

By using (5.18) and (5.19), the right-hand side of (5.18) becomes

$$
\begin{aligned}
\dot{V}(t) \leq\ & \dot{\xi}^T(t)P\xi(t)+\xi^T(t)P\dot{\xi}(t)+\xi^T(t)Q\xi(t)\\
&-\xi^T(t-h)Q\xi(t-h)+h^2\dot{\xi}^T(t)N\dot{\xi}(t)\\
&-\begin{bmatrix}\xi(t-\tau(t))-\xi(t)\\ \xi(t-\tau(t))-\xi(t-h)\end{bmatrix}^T\begin{bmatrix}N & Z\\ Z^T & N\end{bmatrix}\begin{bmatrix}\xi(t-\tau(t))-\xi(t)\\ \xi(t-\tau(t))-\xi(t-h)\end{bmatrix}
\end{aligned}\tag{5.20}
$$

Next, define the following equation:

$$
\mathscr{L}(t)=\dot{V}(t)+\bar{r}^T(t)\bar{r}(t)-\lambda^2w^T(t)w(t)\tag{5.21}
$$

By substituting (5.3) and (5.20) into (5.21) and performing some manipulations, we have

$$
\begin{aligned}
\mathscr{L}(t)=\ & \dot{V}(t)+\bar{r}^T(t)\bar{r}(t)-\lambda^2w^T(t)w(t)\\
\leq\ & \dot{\xi}^T(t)P\xi(t)+\xi^T(t)P\dot{\xi}(t)+\xi^T(t)Q\xi(t)\\
&-\xi^T(t-h)Q\xi(t-h)+h^2\dot{\xi}^T(t)N\dot{\xi}(t)\\
&-\begin{bmatrix}\xi(t-\tau(t))-\xi(t)\\ \xi(t-\tau(t))-\xi(t-h)\end{bmatrix}^T\begin{bmatrix}N & Z\\ Z^T & N\end{bmatrix}\begin{bmatrix}\xi(t-\tau(t))-\xi(t)\\ \xi(t-\tau(t))-\xi(t-h)\end{bmatrix}\\
&+\bar{r}^T(t)\bar{r}(t)-\lambda^2w^T(t)w(t)-e^T(t)\Omega e(t)\\
&+\sigma(y(t-\tau(t))+e(t))^T\Omega(y(t-\tau(t))+e(t))\\
=\ & \sum_{i=1}^{r}\sum_{j=1}^{r}\sum_{l=1}^{r}\mu_i(x)\mu_j(\hat{x})h_l(\hat{x}(t))\varepsilon^T(t)\bar{\Pi}_{ijl}\varepsilon(t)
\end{aligned}\tag{5.22}
$$

where

$$
\varepsilon(t)=[\xi^T(t),\xi^T(t-\tau(t)),\xi^T(t-h),e^T(t),w^T(t)]^T
$$

Note that $\bar{\Pi}_{ijl}<0$ can ensure $\mathscr{L}(t)<0$, and for $\bar{\Pi}_{ijl}$ using the Schur complement formula and congruence transformation, one can obtain Π_{ijl}, which has been defined in Theorem 5.1.

Next, by applying the constraints of the fuzzy weighting parameters and the membership functions, H_∞ performance analysis results are derived, in which the fuzzy weighting parameters and membership functions have the following equation and inequality relationships:

$$0 \leq \mu_j(\hat{x}) \leq 1, \quad \sum_{j=1}^{r} \mu_j(\hat{x}) = 1 \tag{5.23}$$

$$0 \leq h_j(\hat{x}) \leq 1, \quad \sum_{j=1}^{r} h_j(\hat{x}) = 1 \tag{5.24}$$

$$|h_i(\hat{x}) - h_j(\hat{x})| = \alpha|\mu_i(\hat{x}) - \mu_j(\hat{x})| \leq \alpha \leq 1 \tag{5.25}$$

$$0 \leq \alpha \leq 1, \quad \alpha + r\beta = 1, \quad for\ i, j, l = 1, 2, \ldots, r \tag{5.26}$$

Based on the Eq. (5.25), one has

$$\begin{aligned}(\mu_j(x) - \mu_l(x))(h_j(\hat{x}) - h_l(\hat{x})) &= \alpha(\mu_j(x) - \mu_l(x))(\mu_j(\hat{x}) - \mu_l(\hat{x})) \\ &\leq \alpha \end{aligned} \tag{5.27}$$

It should be noted that (5.27) is obtained by multiplying the differences between each membership function in the residual generator and the differences between each membership function in the system, and it will have an important component in the derivation of the H_∞ performance criteria later.

In addition, a slack matrix is introduced, as follows:

$$\sum_{i=1}^{r}\sum_{j=1}^{r}\sum_{l=1}^{r} \mu_i(x)\mu_j(\hat{x})(\mu_l(\hat{x}) - h_l(\hat{x}))K = 0$$

And then, one has

$$\begin{aligned}&\sum_{i=1}^{r}\sum_{j=1}^{r}\sum_{l=1}^{r} \mu_i(x)\mu_j(\hat{x})h_l(\hat{x})\Pi_{ijl} + \sum_{i=1}^{r}\sum_{j=1}^{r}\sum_{l=1}^{r} \mu_i(x)\mu_j(\hat{x})(\mu_l(\hat{x}) - h_l(\hat{x}))K \\
&= \sum_{i=1}^{r}\sum_{j=1}^{r}\sum_{l=1}^{r} \mu_i(x)\mu_j(\hat{x})(\mu_l(\hat{x})(\alpha\Pi_{ijl} - \alpha K + K) \\
&\quad + (h_l(\hat{x}) - \alpha\mu_l(\hat{x}))(\Pi_{ijl} - K)) \\
&= \frac{1}{2}\sum_{i=1}^{r}\sum_{j=1}^{r}\sum_{l=1}^{r} \mu_i(x)\mu_j(\hat{x})\mu_l(\hat{x})(\alpha\Pi_{ijl} + \alpha\Pi_{ilj} - 2\alpha K + 2K) \\
&\quad + \sum_{i=1}^{r}\sum_{j=1}^{r}\sum_{l=1}^{r} \mu_i(x)\mu_j(\hat{x})(h_l(\hat{x}) - \alpha\mu_l(\hat{x}))(\Pi_{ijl} - K) \\
&< \frac{1}{2}\sum_{i=1}^{r}\sum_{j=1}^{r}\sum_{l=1}^{r} \mu_i(x)\mu_j(\hat{x})\mu_l(\hat{x})(\alpha\Pi_{ijl} + \alpha\Pi_{ilj} - 2\alpha K + 2K)\end{aligned} \tag{5.28}$$

where the inequality $\Pi_{ijl} - K < 0$ and $h_l(\hat{x}) - \alpha\mu_l(\hat{x}) = \beta > 0$ is used in the second step.

Next, according to the constraints (5.24), (5.25) and (5.27), the following equation and inequalities are established:

$$-\sum_{i=1}^{r}\mu_i(x)\left(\sum_{j=1}^{r}\sum_{l=1}^{r}\mu_j(\hat{x})\mu_l(\hat{x}) - \sum_{j=1}^{r}\mu_j(\hat{x}) - \sum_{l=1}^{r}\mu_l(\hat{x}) + 1\right) = 0 \tag{5.29}$$

$$\sum_{i=1}^{r}\mu_i(x)\sum_{j=1}^{r-1}\sum_{l>j}^{r}(\alpha - \mu_j(\hat{x})\mu_j(\hat{x}) + \mu_j(\hat{x})\mu_l(\hat{x}) + \mu_l(\hat{x})\mu_j(\hat{x}) - \mu_l(\hat{x})\mu_l(\hat{x})) \geq 0 \tag{5.30}$$

$$-\sum_{i=1}^{r}\mu_i(x)\sum_{j=1}^{r}\mu_j(\hat{x})(\mu_j(x) - 1) \geq 0 \tag{5.31}$$

$$-\sum_{i=1}^{r}\mu_i(x)\sum_{j=1}^{r}\mu_j(\hat{x})(\mu_j(\hat{x}) - 1) \geq 0 \tag{5.32}$$

Substituting (5.29)–(5.32) with $\Lambda_a > 0$, $\Lambda_b > 0$ and $\Lambda_i > 0$ $(i = 1, 2, \ldots, r)$ into the inequality (5.28), yields:

$$\begin{aligned}
&\frac{1}{2}\sum_{i=1}^{r}\sum_{j=1}^{r}\sum_{l=1}^{r}\mu_i(x)\mu_j(\hat{x})\mu_l(\hat{x})(\alpha\Pi_{ijl} + \alpha\Pi_{ilj} - 2\alpha K + 2K)\\
&< \frac{1}{2}\sum_{i=1}^{r}\sum_{j=1}^{r}\sum_{l=1}^{r}\mu_i(x)\mu_j(\hat{x})\mu_l(\hat{x})(\alpha\Pi_{ijl} + \alpha\Pi_{ilj} - 2\alpha K + 2K)\\
&\quad - \Lambda_a\sum_{i=1}^{r}\mu_i(x)(\sum_{j=1}^{r}\sum_{l=1}^{r}\mu_j(\hat{x})\mu_l(\hat{x}) - \sum_{j=1}^{r}\mu_j(\hat{x}) - \sum_{l=1}^{r}\mu_l(\hat{x}) + 1)\\
&\quad + \Lambda_b\sum_{i=1}^{r}\mu_i(x)\sum_{j=1}^{r-1}\sum_{l>j}^{r}(\alpha - \mu_j(\hat{x})\mu_j(\hat{x})\\
&\quad + \mu_j(\hat{x})\mu_l(\hat{x}) + \mu_l(\hat{x})\mu_j(\hat{x}) - \mu_l(\hat{x})\mu_l(\hat{x}))\\
&\quad - \sum_{i=1}^{r}\mu_i(x)\sum_{j=1}^{r}\Lambda_j(\mu_j(\hat{x})\mu_j(x) - \mu_j(\hat{x}))\\
&\quad - \sum_{i=1}^{r}\mu_i(x)\sum_{j=1}^{r}\Lambda_j(\mu_j(\hat{x})\mu_j(\hat{x}) - \mu_j(\hat{x}))
\end{aligned} \tag{5.33}$$

Then, the above equation is reconstructed by using membership functions $\mu_j(\hat{x})$ and $\mu_l(\hat{x})$ and described below:

$$\sum_{i=1}^{r} \mu_i(x) \begin{bmatrix} \Theta \times I_{2n} \\ I_{2n} \end{bmatrix}^T X_i \begin{bmatrix} \Theta \times I_{2n} \\ I_{2n} \end{bmatrix} < 0 \quad (5.34)$$

where $\Theta = \left[\mu_1^T(\hat{x}) \ \mu_2^T(\hat{x}) \ \cdots \ \mu_r^T(\hat{x}) \right]^T$.

Consequently, for the given scale parameter $0 \le \alpha \le 1$ and offset parameter $\beta = \frac{1-\alpha}{r}$, if $X_i < 0$ ($i = 1, 2, \ldots, r$) are satisfied, then (5.12) will satisfy the inequality condition (5.13). The proof is completed. □

5.3.2 Fault Detection Observer Design Conditions

The matrix inequalities presented in Theorem 5.1 are coupled and difficult for the design of the residual generators and fault detection observers. Therefore, next, by selecting an appropriate matrix P, the inequalities in Theorem 5.1 will be linearized, such that it is convenient for the design of the fault detection observers.

Theorem 5.2 *For the given parameters $h > 0$, $0 \le \sigma < 1$, $0 \le \alpha \le 1$ and $\beta = \frac{1-\alpha}{r}$. The fault detection system (5.12) satisfies (Q.1) and (Q.2), if there exist matrices $P = \begin{bmatrix} Y_1 & -Y_2 \\ -Y_2 & Y_2 \end{bmatrix} > 0$, $Q > 0$, $N > 0$, $\Lambda_a > 0$, $\Lambda_b > 0$, $\Lambda_i > 0$, $i = 1, 2, \ldots, r$, Z, K, $\mathscr{A}_j$, and $\lambda > 0$, the following inequalities are established for $i = 1, 2, \ldots, r$:*

$$\hat{\Pi}_{ijl} - K < 0 \quad (5.35)$$

$$\begin{bmatrix} N & Z \\ Z^T & N \end{bmatrix} \ge 0 \quad (5.36)$$

$$\hat{X}_i = \begin{bmatrix} \hat{X}_{i11} & \hat{X}_{i12} & \cdots & \hat{X}_{i1r} & \Gamma_{i1} \\ \star & \hat{X}_{i22} & \cdots & \hat{X}_{i2r} & \Gamma_{i2} \\ \star & \star & \ddots & \vdots & \vdots \\ \star & \star & \star & \hat{X}_{irr} & \Gamma_{ir} \\ \star & \star & \star & \star & \Upsilon_i \end{bmatrix} < 0 \quad (5.37)$$

where the parameters $\hat{X}_i$ and $\hat{\Pi}_{ijl}$ are obtained by replacing P in the corresponding matrices in Theorem 5.2 with $\begin{bmatrix} Y_1 & -Y_2 \\ -Y_2 & Y_2 \end{bmatrix}$.

Then, the gains of the fault detection observer are reconstructed as $L_j = (\mathscr{A}_j Y_2^{-1})^T$, and the gains of the residual generator and weighted matrix are obtained directly.

Proof Let the matrix $P = \begin{bmatrix} Y_1 & -Y_2 \\ -Y_2 & Y_2 \end{bmatrix}$, define $L_j^T Y_2$ as a new variable $\mathscr{A}_j$. Then, the matrices in Theorem 5.1 are linearized. The proof is completed.

Next, the relevant parameters are solved through an algorithm as follows.

Algorithm 1: By computing (5.37), (5.35) and (5.36), the parameters of the fault detection observer, the gains of the residual generator and weighted matrix are computed through the following optimisation:

$$\begin{aligned} &\min \lambda \\ &s.t.\ (5.37), (5.35), (5.36) \end{aligned} \tag{5.38}$$

Remark 5.3 In Ref. [17], the membership function rules are exchanged among different membership functions, which is obviously not tenable. Therefore, there are some limitations and defects in the existing methods. The method proposed in this chapter effectively solves the design of event-triggered fault detection observers, while overcoming the shortcomings of the existing results.

5.4 Numeral Example

In this section, the proposed policy is implemented in the fault detection of tunnel diode circuit in Refs. [4, 7]. The tunnel diode circuit is shown in Figure 2.1 of Chap. 2 and is modeled as follows:

$$i_D(t) = 0.002 v_D(t) + 0.01 v_D^3(t)$$

Similar to the Ref. [4], define $v_C(t)$ and $i_L(t)$ as the state variables $x_1(t)$ and $x_2(t)$, respectively. Then, the state equation of the circuit is established below:

$$\begin{aligned} C\dot{x}_1(t) &= -0.002 x_1(t) - 0.01 x_1^3(t) + x_2(t) \\ L\dot{x}_2(t) &= -x_1(t) - R x_2(t) \end{aligned}$$

where the parameters $C = 20\,\text{mF}$, $L = 1000\,\text{mH}$ and $R = 10\,\Omega$. In order to obtain fewer fuzzy rules, the local approximation method in fuzzy partition spaces is chosen to establish the T-S fuzzy system (5.1), in which the membership functions are plotted in Fig. 5.1. Select an appropriate output matrix, and then the corresponding parameters are given as follows:

$$\begin{aligned} A_1 &= \begin{bmatrix} -0.1 & 50 \\ -1 & -10 \end{bmatrix}, A_2 = \begin{bmatrix} -4.6 & 50 \\ -1 & -10 \end{bmatrix}, D_1 = \begin{bmatrix} 0 \\ 1 \end{bmatrix}, \\ D_2 &= \begin{bmatrix} 0 \\ 1 \end{bmatrix}, F_1 = \begin{bmatrix} 1 \\ 0 \end{bmatrix}, F_2 = \begin{bmatrix} 1 \\ 0 \end{bmatrix}, C = \begin{bmatrix} 1 & 0 \\ 0 & 1 \end{bmatrix}. \end{aligned}$$

The membership functions in residual generator match those in T-S fuzzy system, and moreover, since the states of the T-S fuzzy system are unknown, the functions of

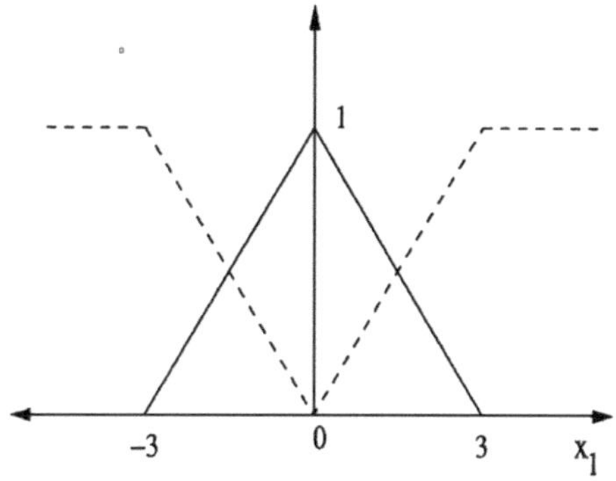

Fig. 5.1 Membership functions

system states and observer states are, respectively, selected as the premise variables of the T-S fuzzy system, fault detection observer and residual generator.

The purpose is to construct a matching membership function approach to overcome the shortcomings of the existing results, and to realize the fault detection of the T-S fuzzy system at the same time. In order to prove the effectiveness of the constructed approach, two sets of scale and offset parameters are selected and discussed in two cases.

Case 1: Scale coefficient $\alpha = 0.4$, offset coefficient $\beta = 0.3$.

By solving the linear matrix inequalities in Theorem 5.2, the gains of the fault detection observer and residual generator, and the weighted matrix are obtained as follows:

$$L_1 = \begin{bmatrix} 26.5226 & 1.2882 \\ 1.2882 & 29.1844 \end{bmatrix}, L_2 = \begin{bmatrix} 3.4909 & 8.5603 \\ -0.5284 & -1.8088 \end{bmatrix},$$
$$W_1 = \begin{bmatrix} -0.0062 & -0.0108 \\ -0.0062 & -0.0108 \end{bmatrix}, W_2 = \begin{bmatrix} -0.0006 & 0.0011 \\ -0.0006 & 0.0011 \end{bmatrix},$$
$$\Omega = \begin{bmatrix} 26.5226 & 1.2882 \\ 1.2882 & 29.1844 \end{bmatrix}.$$

For demonstration, the disturbance is selected as $v(t) = 0.4sin(0.314t)$ $(0 \le t \le 30)$, and similar to the Refs. [19, 20], suppose that there is a bias fault in the capacitor voltage, and it is set to $f(t) = 1.5 + 1.5sin(0.1t)$ $(15 \le t \le 30)$.

The main results are shown in Figs. 5.2, 5.3, 5.4, 5.5 and 5.6 by using the solved gains and considering the external disturbances and faults. The release instants and release intervals of the event-triggered mechanism are shown in Fig. 5.2, and the outputs of the T-S fuzzy system and the inputs of the fault detection observer are plotted in Figs. 5.3 and 5.4. It can be seen from Figs. 5.2 to 5.4 that the system outputs are only transmitted at the triggering instants, and the number of measurement outputs is 600, and the data to be transmitted is 293. Therefore, the introduction of event-triggered mechanism can effectively save network resources. The residual signals are

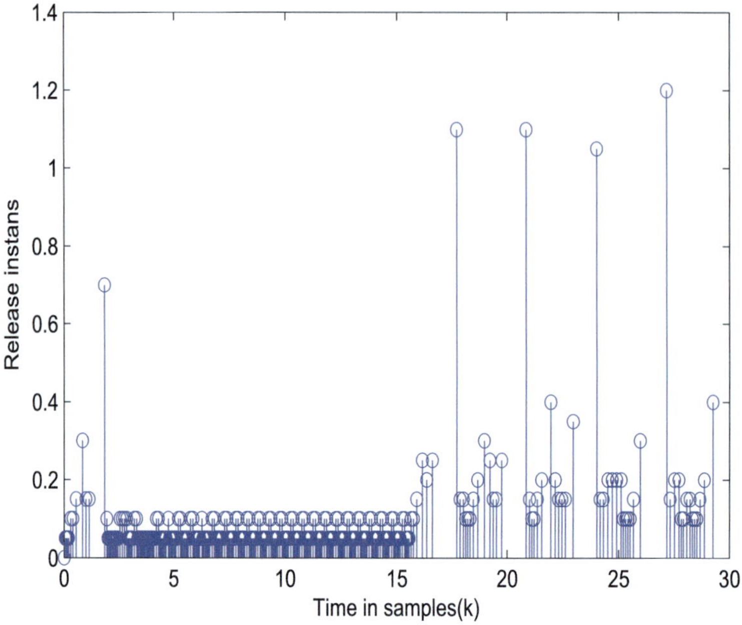

Fig. 5.2 The release instants and time intervals

plotted in Fig. 5.5, where the residual signals jump significantly when the fault occurs, which indicates that the residual signals are sensitive to the fault. The evaluation function and threshold are presented in Fig. 5.6, from which it can be seen that the fault is detected at $k = 17.59$.

Case 2: Scale coefficient $\alpha = 0.2$, offset coefficient $\beta = 0.4$.

In this case, the same parameters as in Case 1 are selected except for the scale coefficient α and offset coefficient β, and by solving the linear matrix inequalities in Theorem 5.2, the gains of the fault detection observer and residual generator, and the weighted matrix are obtained as follows:

$$
\begin{aligned}
L_1 &= \begin{bmatrix} 3.5823 & 8.8380 \\ -0.5475 & -1.8554 \end{bmatrix}, L_2 = \begin{bmatrix} -0.5303 & 1.6827 \\ 0.0819 & -0.5712 \end{bmatrix}, \\
W_1 &= \begin{bmatrix} -0.0069 & -0.0103 \\ -0.0069 & -0.0103 \end{bmatrix}, W_2 = \begin{bmatrix} -0.0016 & 0.0002 \\ -0.0016 & 0.0002 \end{bmatrix}, \\
\Omega &= \begin{bmatrix} 27.1921 & 1.2225 \\ 1.2225 & 29.6833 \end{bmatrix}.
\end{aligned}
$$

In addition, the same disturbance and fault as in Case 1 are also selected. Then, based on the existing parameters, the residuals, the evaluation function and threshold are

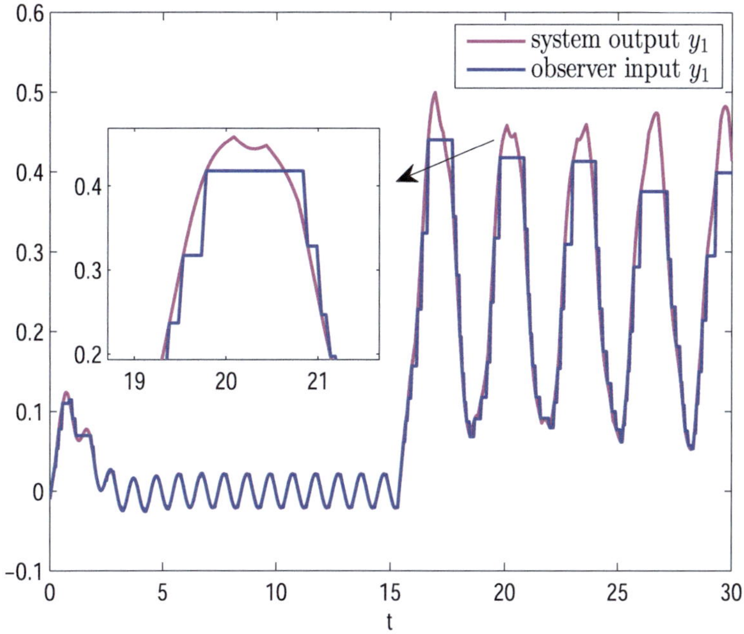

Fig. 5.3 System output y_1 and observer input y_1

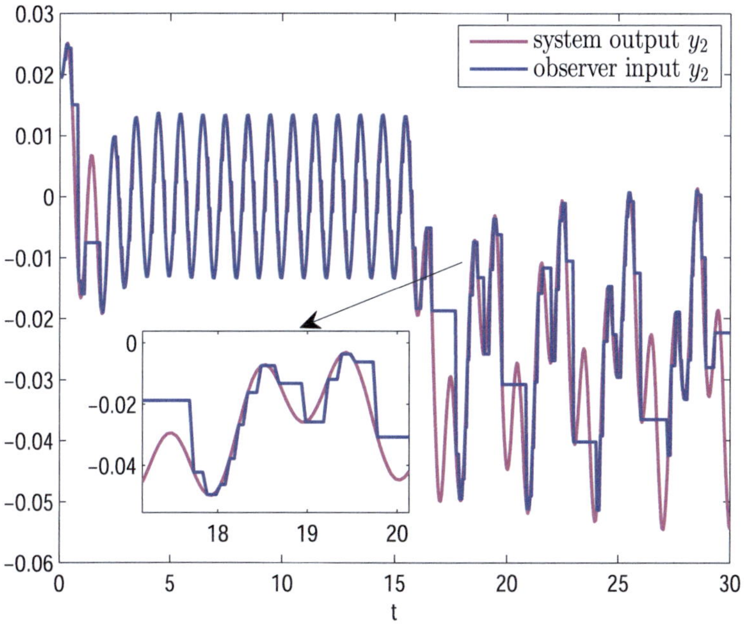

Fig. 5.4 System output y_2 and observer input y_2

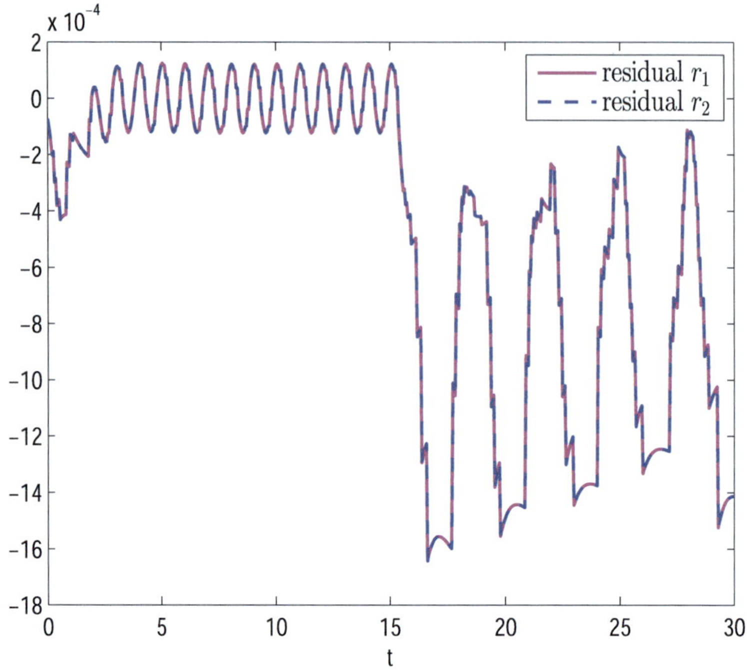

Fig. 5.5 Residual signals

shown in Figs. 5.7 and 5.8 respectively. Figure 5.7 shows that the residual signals are sensitive to the fault, and Fig. 5.8 shows that the fault is detected at $k = 17.43$.

In summary, the event-triggered fault detection scheme proposed in this chapter is effective.

5.5 Summary

This chapter has studied the problem of event-triggered fault detection for continuous-time T-S fuzzy systems. First, an event-triggered fuzzy fault detection observer is designed for T-S fuzzy system, and a residual generator with matching membership functions is constructed to generate residual signals. Subsequently, by applying the constraints of the fuzzy weighting parameters and the differences of each membership functions, new criteria in terms of linear matrix inequalities are derived to ensure the desired performance of the fault detection system. Finally, an example is shown to illustrate the validity of the designed fault detection scheme.

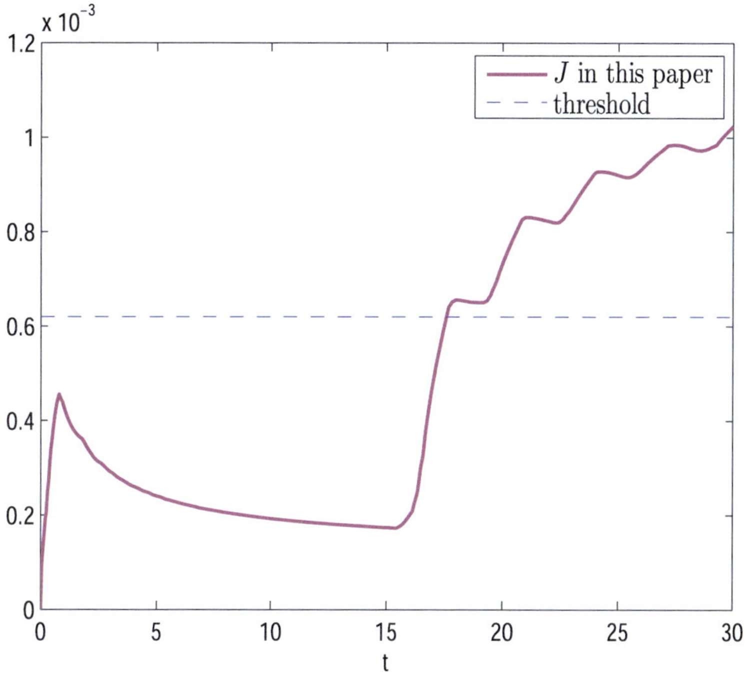

Fig. 5.6 Residual evaluation function and threshold

5.6 Notes

Fault detection of T-S fuzzy systems in networked environments is crucial for improving the safety and reliability of practical systems. Meanwhile, the introduction of event-triggered mechanisms can save network resources and also cause the asynchronous of the premise variables, which poses difficulties in the design of fault detection observers. Therefore, it is necessary to construct a new fault detection strategy to address the mismatch of premise variables in response to the shortcomings of the existing methods in Refs. [13–18], which motivates this paper. In addition, it should be noted that the results presented in this chapter are mainly constructed based on the ones in Ref. [21].

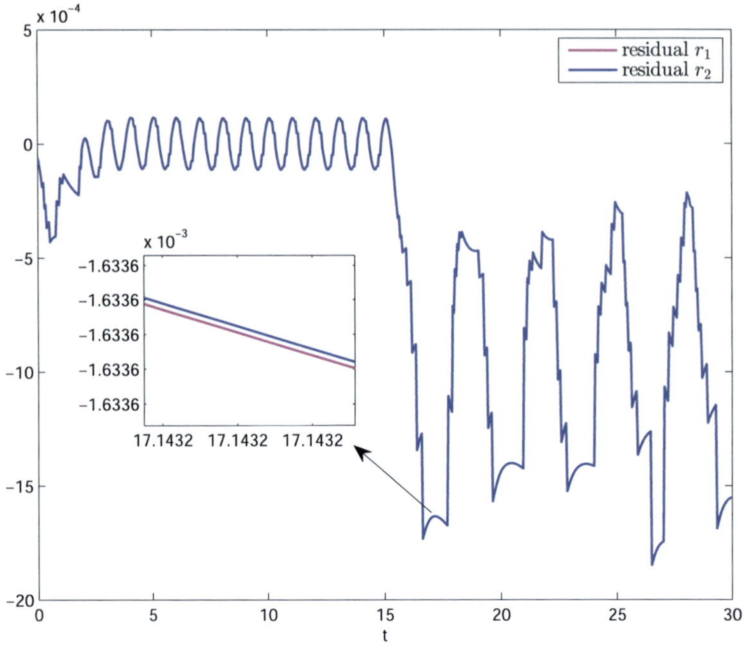

Fig. 5.7 Residual signals

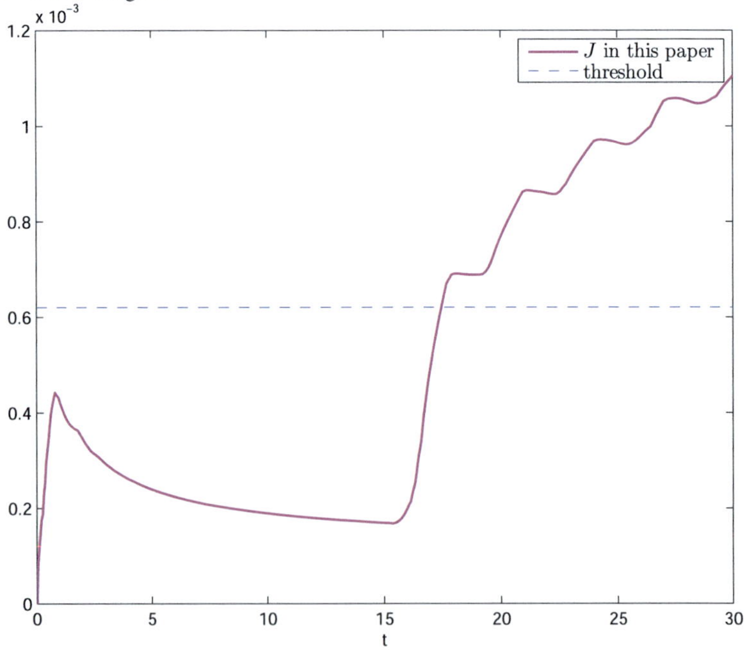

Fig. 5.8 Residual evaluation function and threshold

References

1. Liu M, Zhang LX, Shi P, Zhao YX (2018) Fault estimation sliding-mode observer with digital communication constraints. IEEE Trans Autom Control 63(10):3434–3441
2. Chen H, Jiang B, Ding SX, Huang B (2022) Data-driven fault diagnosis for traction systems in high-speed trains: a survey, challenges, and perspectives. IEEE Trans Intell Transp Syst 23(3):1700–1716
3. Liu MJ, Luo P, Hu CH, Guo R, Hu XX (2023) Actuator fault detection of T-S fuzzy hypersonic flight vehicle model: A T-S fuzzy-based H_∞ sliding mode observer approach. IEEE J Miniaturization Air Space Syst 4(3):274–282
4. Wang XL, Yang GH (2021) Event-triggered fault detection observer design for T-S fuzzy systems. IEEE Trans Fuzzy Syst 29(9):2532–2542
5. Wang XL, Yang GH (2020) Distributed event-triggered H_∞ filtering for discrete-time T-S fuzzy systems over sensor networks. IEEE Trans Syst Man Cybern: Syst 50(9):3269–3280
6. Li HY, Chen ZR, Wu LG, Lam HK, Du HP (2017) Event-triggered fault detection of nonlinear networked systems. IEEE Trans Cybern 47(4):1041–1052
7. Pan YN, Yang GH (2018) Event-triggered fault detection filter design for nonlinear networked systems. IEEE Trans Syst Man Cybern 48(11):1851–1862
8. Yang HL, Wang X, Park JH (2022) Sampled-data-based dissipative stabilization of IT-2 TSFSs via fuzzy adaptive event-triggered protocol. IEEE Trans Cybern 52(22):11594–11603
9. Guo XG, Fan X, Ahn CK (2021) Adaptive event-triggered fault detection for interval Type-2 T-S fuzzy systems with sensor saturation. IEEE Trans Fuzzy Syst 29(8):2310–2321
10. Zhang Q, Yan HC, Wang M, Li ZC, Chang YF (2023) Asynchronous fault detection filter design for T-S fuzzy singular systems via dynamic event-triggered scheme. IEEE Trans Fuzzy Syst 31(3):970–981
11. Gu Z, Yue D, Park JH, Xie XP (2023) Memory-event-triggered fault detection of networked IT2 T-S fuzzy systems. IEEE Trans Cybern 53(2):743–752
12. Zhang CZ, Hu JF, Qiu JB, Chen QJ (2017) Event-triggered nonsynchronized H_∞ filtering for discrete-time T-S fuzzy systems based on piecewise Lyapunov functions. IEEE Trans Syst Man Cybern: Syst 47(8):2330–2341
13. Lam HK, Narimani M (2009) Stability analysis and performance design for fuzzy-model-based control system under imperfect premise matching. IEEE Trans Fuzzy Syst 17(4):949–961
14. Peng C, Ma SD, Xie XP (2017) Observer-based non-PDC control for networked T-S fuzzy systems with an event-triggered communication. IEEE Trans Cybern 47(8):2279–2287
15. Lam HK, Leung FHF (2007) Sampled-data fuzzy controller for time-delay nonlinear systems: fuzzy-model-based LMI approach. IEEE Trans Syst Man Cybern, Part B (Cybern) 37(3):617–629
16. Li HY, Wu C, Yin S, Lam HK (2016) Observer-based fuzzy control for nonlinear networked systems under unmeasurable premise variables. IEEE Trans Fuzzy Syst 24(5):1233–1245
17. Jin YS, Kwon WY, Lee S (2020) Further results on sampled-data H_∞ filtering for T-S fuzzy systems with asynchronous premise variables. IEEE Trans Fuzzy Syst 30(6):1864–1874
18. Lee S (2019) Novel stabilization criteria for T-S fuzzy systems with affine matched membership functions. IEEE Trans Fuzzy Syst 27(3):540–548
19. Zhao Y, Lam J, Gao HJ (2009) Fault detection for fuzzy systems with intermittent measurements. IEEE Trans Fuzzy Syst 17(2):398–410
20. Wu CW, Li HY, Lam HK, Karimi HR (2016) Fault detection for nonlinear networked systems based on quantization and dropout compensation: an interval type-2 fuzzy-model method. Neurocomputing 191:409–420
21. Wang XL, Hao LY (2022) Event-triggered fault detection for Takagi-Sugeno fuzzy systems via an improved matching membership function approach. Inform Sci 593:35–48

Chapter 6
Event-Triggered Switched Fault Detection Observer Design for T-S Fuzzy Systems

6.1 Introduction

The design of event-triggered fault detection observers has been discussed in this chapter, and several existing methods for dealing with the asynchronous problems of premise variables in T-S fuzzy systems in networked environments have also been introduced, please refer to Refs. [1–6], including partition method, non-PDC method, etc. At the same time, it is pointed out that there are limitations and defects in these types of methods, and a new matching membership function strategy is proposed to handle the mismatch of premise variables in view of the shortcomings of the existing non-PDC methods, thereby solving the design of event-triggered fault detection observers.

It should be noted that in addition to the methods of dealing with the asynchronous problems of premise variables introduced in Chap. 6, an asynchronous premise reconstruction method was proposed in Refs. [7, 8] to deal with the mismatch of premise variables. However, it is important to note that the asynchronous premise reconstruction methods proposed in Refs. [7, 8] require the upper and lower bounds of λ to be given before the controllers are designed. In fact, it is known from (13) to (17) in Ref. [7] that the upper and lower bounds of λ cannot be computed before the controllers are calculated. In this case, the results in Refs. [7, 8] will not be able to perfectly solve the mismatch problem of premise variables. There, it is necessary to construct a new asynchronous premise reconstruction method to improve the defect in this method. Moreover, note that in the event-triggered results mentioned above, the triggering instants are not effectively utilized, which will lead to a certain degree of conservatism. Therefore, how to effectively use the triggering instants to establish better fault detection mechanisms is an important and meaningful problem, which motivates this chapter.

This chapter is concerned with the problem of event-triggered fault detection observer design for continuous-time T-S fuzzy systems. The nonlinear systems considered are presented as T-S fuzzy systems, and the system outputs are selected as the

X.-L. Wang et al., *Robust Filtering and Fault Detection for T-S Fuzzy Systems*,
https://doi.org/10.1007/978-981-96-5818-3_6

premise variables of T-S fuzzy systems. New types of observers, namely, switched fuzzy fault detection observers, are designed, which are switched at different triggering instants based on the online asynchronous premise reconstruction method. By employing new triggering instants-dependent Lyapunov functions established via the idea of event interval partitioning, it is proved that the proposed fault detection observers can guarantee the stability and desired performance of the fault detection systems. Finally, a simulation example is applied to show that the proposed fault detection scheme can achieve better fault detection results than the existing design methods. At the same time, the online asynchronous premise reconstruction method designed in this chapter can perfectly solve the asynchronous problem of premise variables.

6.2 System Modeling and Problem Statement

A class of continuous-time T-S fuzzy systems is considered, as shown below:

$$\begin{aligned}\dot{x}(t) &= \sum_{i=1}^{r} \mu_i(\zeta(y(t)))(A_i x(t) + D_i v(t) + G_i f(t)) \\ y(t) &= Cx(k)\end{aligned} \tag{6.1}$$

where r is the number of inference rules. $\zeta_\iota(y(t))$, $\iota = 1, \ldots, p$ are the premise variables that depend on the measurable outputs $y(t) \in R^m$. $v(t) \in R^v$ and $f(t) \in R^f$ are, respectively, the external disturbances and actuator faults belonging to L_2. $x(t) \in R^n$ are the system states. $A_i \in R^{n\times n}$, $D_i \in R^{n\times v}$, $G_i \in R^{n\times f}$ and $C \in R^{m\times n}$ are known constant matrices.

In addition, membership function is defined as $\mu_i(\zeta(y(t)))$ and described as follows:

$$\sum_{i=1}^{r} \mu_i(\zeta(y(t))) = 1,\ \mu_i(\zeta(y(t))) \geq 0,\ \forall i = 1, 2, \ldots, r \tag{6.2}$$

The event-triggered mechanism is introduced in this chapter to save network resources. First, it is assumed that the measurement outputs transmitted at the triggering instant $t_{q-1}h$ $(q = 1, 2, \ldots, N)$ are $y(t_{q-1}h)$, where $t_0 = 0$, and the first event is generated at the triggering instant $t_0 h$. The transmission of the current signals is determined by comparing the measurement errors between outputs $y(t)$ and signals $y(t_{q-1}h)$ with the threshold dependent on the outputs. Then, the next transmission instant is described below [2, 11]:

$$t_q h = t_{q-1}h + \min_{n\in N}\{nh | y_s^T \Phi y_s > \sigma y^T(t_{q-1}h)\Phi y(t_{q-1}h)\} \tag{6.3}$$

in which $y_s = y(t_{q-1}h) - y(i_{q-1}h)$, $i_{q-1}h = t_{q-1}h + nh$, h is the sampling period. The weight matrix and threshold are defined as $\Phi \in R^{m\times m}$ and σ respectively, and satisfy $\Phi > 0$ and $\sigma \in [0,\ 1)$. Moreover, one knows that $\{t_0, t_1, t_2, \ldots\} \subset \{0, 1, 2, \ldots\}$.

The sampled data are transmitted between the event unit and the ZOHs via a network channel that is supposed to be ideal such that the sampled signal can immediately reach the ZOHs and be held unchanged until the next sampled data arrives. Thus, the following equation is established:

$$\bar{y}(t) = y(t_{q-1}h), \quad t \in [t_{q-1}h, t_q h) \tag{6.4}$$

Next, first define $[t_{q-1}h, t_q h) = \bigcup_{n=0}^{\varrho_{q-1}} \aleph_n$, where $\aleph_n = [t_{q-1}h + nh, t_{q-1}h + nh + h)$, $n = 0, 1, \ldots, \varrho_{q-1}$, and $\varrho_{q-1} = t_q - t_{q-1} - 1$. Then, introduce the new variables $\tau(t)$, $e(t)$, and describe them as follows:

$$\tau(t) = \begin{cases} t - t_{q-1}h, \ t \in \aleph_0, \\ t - t_{q-1}h - h, \ t \in \aleph_1, \\ \quad\vdots \\ t - t_{q-1}h - \varrho_{q-1}h, \ t \in \aleph_{\varrho_{q-1}} \end{cases} \tag{6.5}$$

$$e(t) = \begin{cases} y(t_{q-1}h) - y(t_{q-1}h), \ t \in \aleph_0, \\ y(t_{q-1}h) - y(t_{q-1}h + h), \ t \in \aleph_1, \\ \quad\vdots \\ y(t_{q-1}h) - y(t_{q-1}h + \varrho_{q-1}h), \ t \in \aleph_{\varrho_{q-1}} \end{cases} \tag{6.6}$$

According to (6.5), one obtains:

$$0 \le \tau(t) \le h \tag{6.7}$$

Furthermore, similar to the methods in Refs. [9, 10], the following equation is established according to (6.5) and (6.6):

$$y(t_{q-1}h) = y(t - \tau(t)) + e(t) \tag{6.8}$$

Based on the above analysis, it is known that the measurement outputs $y(t_{q-1}h)$ are transmitted to the observer side and applied to establish the fault detection observers. Next, new types of observers, namely, event-triggered fault detection observers, are established to detect the fault signals. The structure of the constructed event-triggered fuzzy fault detection observers is described as follows.

If $\zeta_1(y(t_{q-1}h))$ is υ_{j1} and $\zeta_2(y(t_{q-1}h))$ is υ_{j2} and $\cdots$ and $\zeta_p(y(t_{q-1}h))$ is υ_{jp}, then

$$\dot{\hat{x}}(t) = \begin{cases} A_j\hat{x}(t) + L_{j,1}(y(t_0h) - \hat{y}(t)),\ t \in [t_0h,\ t_1h) \\ \vdots \\ A_j\hat{x}(t) + L_{j,q}(y(t_{q-1}h) - \hat{y}(t)),\ t \in [t_{q-1}h,\ t_qh) \\ \vdots \\ A_j\hat{x}(t) + L_{j,N}(y(t_{N-1}h) - \hat{y}(t)),\ t \in [t_{N-1}h,\ t_Nh) \end{cases}$$
$$\hat{y}(t) = C\hat{x}(t)$$
$$r(t) = y(t_{q-1}h) - \hat{y}(t),\ t \in [t_{q-1}h,\ t_qh) \tag{6.9}$$

where $L_{j,q} \in R^{n\times m}$, $q = 1, 2, \ldots, N$ are the gains of fault detection observers.

It is noted that the measurement outputs may not be accurately received by the fault detection observers due to the existence of event-triggered mechanism, which results in the mismatch of premise variables. Therefore, the membership functions of the fault detection observers need to be reconstructed in the design of the fault detection observers. Based on this view, an online asynchronous premise reconstruction approach is proposed and the detailed reconstruction steps are given as follows:

First, the time-varying upper bound of the difference between asynchronous membership functions is established, as shown below:

$$|\mu_j(\zeta(y(t))) - \mu_j(\zeta(y(t_{q-1}h)))| \leq \Upsilon_{j,q} \tag{6.10}$$

in which $\Upsilon_{j,q} \geq 0$. Then, one can construct the proportional relationship between asynchronous membership functions:

$$\mu_j(\zeta(y(t_{q-1}h))) = \Delta_{j,q}(\zeta(y(t)), \zeta(y(t_{q-1}h)))\mu_j(\zeta(y(t))) \tag{6.11}$$

where $\Delta_{j,q}(\zeta(y(t)), \zeta(y(t_{q-1}h)))$ is the proportional coefficient.

Furthermore, combine (6.10) and (6.11) to get:

$$\begin{aligned} \Theta_{1j,q} &= 1 - \frac{\Upsilon_{j,q}}{\mu_j(\zeta(y(t)))} \\ &\leq \Delta_{j,q}(\zeta(y(t)), \zeta(y(t_{q-1}h))) \\ &\leq 1 + \frac{\Upsilon_{j,q}}{\mu_j(\zeta(y(t)))} \\ &= \Theta_{2j,q} \end{aligned} \tag{6.12}$$

in which $\Theta_{1j,q}$ and $\Theta_{2j,q}$ denote the minimum and maximum value of $\Delta_{j,q}(\zeta(y(t)), \zeta(y(t_{q-1}h)))$, respectively, and satisfy:

$$\begin{aligned} \frac{\Theta_{1i,q}}{\Theta_{2j,q}} &= \frac{min\{\Delta_{i,q}\}}{max\{\Delta_{j,q}\}} \\ &\leq \frac{\Delta_{i,q}(\zeta(y(t)), \zeta(y(t_{q-1}h)))}{\Delta_{j,q}(\zeta(y(t)), \zeta(y(t_{q-1}h)))} \end{aligned}$$

$$
\begin{aligned}
&\le \frac{max\{\Delta_{i,q}\}}{min\{\Delta_{j,q}\}} \\
&= \frac{\Theta_{2i,q}}{\Theta_{1j,q}}
\end{aligned}
\tag{6.13}
$$

Define $\Theta_{1,q} = min\{\Theta_{1i,q}\}$ and $\Theta_{2,q} = max\{\Theta_{2i,q}\}$, the following inequalities are further established:

$$
\gamma_{1,q} = \frac{\Theta_{1,q}}{\Theta_{2,q}} \le \frac{\Delta_{i,q}(\zeta(y(t)), \zeta(y(t_{q-1}h)))}{\Delta_{j,q}(\zeta(y(t)), \zeta(y(t_{q-1}h)))} \le \frac{\Theta_{2,q}}{\Theta_{1,q}} = \gamma_{2,q} \tag{6.14}
$$

Based on the proposed asynchronous premise reconstruction approach, the event-triggered fault detection observers are established in the interval $[t_{q-1}h,\ t_q h)$ and presented as follows:

$$
\begin{aligned}
\dot{\hat{x}}(t) &= \sum_{j=1}^{r} \Delta_{j,q}(\zeta(y(t)), \zeta(y(t_{q-1}h)))\mu_j(\zeta(y(t))) \\
&\quad \times (A_j\hat{x}(t) + L_{j,q}(y(t-\tau(t)) + e(k) - \hat{y}(t))) \\
\hat{y}(t) &= C\hat{x}(t) \\
r(t) &= y(t_{q-1}h) - \hat{y}(t)
\end{aligned}
\tag{6.15}
$$

Remark 6.1 The asynchronous premise reconstruction method has been investigated in Refs. [7, 8], in which the upper and lower bounds of λ are given before the controllers are designed. However, it is known from (6.12) to (6.16) in Ref. [7] that the upper and lower bounds of λ are not easy to get before the controllers are calculated. In this case, the results in Refs. [7, 8] will not be able to perfectly solve the mismatch of premise variables. Therefore, an online asynchronous premise reconstruction method is designed in this chapter to improve the defect of the existing results.

Define new variables $\xi(t) = \begin{bmatrix} x^T(t) & \tilde{x}^T(t) \end{bmatrix}^T \in R^{2n}$, $\tilde{x}(t) = x(t) - \hat{x}(t) \in R^n$, $w(t) = \begin{bmatrix} v^T(t) & f^T(t) \end{bmatrix}^T \in R^{v+f}$, and $\bar{r}(t) = r(t) - Mf(t) \in R^m$, in which $M = \begin{bmatrix} 1 \cdots 1 \end{bmatrix}^T \in R^{m\times s}$, and then combine the T-S fuzzy system (6.1) with the fault detection observers (6.15), one has

$$
\begin{aligned}
\dot{\xi}(t) &= \sum_{i=1}^{r}\sum_{j=1}^{r} \Delta_{j,q}\mu_i\mu_j(\bar{A}_{ij}^q\xi(t) + \bar{B}_{ij}^q\xi(k-\tau(t)) + \bar{E}_{ij}^q e(t) + \bar{D}_{ij}^q w(t)) \\
\bar{r}(t) &= \sum_{i=1}^{r}\sum_{j=1}^{r} \Delta_{j,q}\mu_i\mu_j(\bar{C}_{ij}^q\xi(t) + \bar{G}_{ij}^q\xi(k-\tau(t)) + \bar{D}_{fj}^q e(t) + \bar{H}w(t))
\end{aligned}
\tag{6.16}
$$

in which the parameter matrices are given as follows:

$$\bar{A}_{ij}^q = \begin{bmatrix} A_i & 0 \\ A_i - A_j + L_j C & A_j - L_j C \end{bmatrix}, \bar{E}_{ij}^q = \begin{bmatrix} 0 \\ -L_j \end{bmatrix}, \bar{B}_{ij}^q = \begin{bmatrix} 0 & 0 \\ -L_j C & 0 \end{bmatrix},$$
$$\bar{D}_{ij}^q = \begin{bmatrix} D_i & G_i \\ D_i & G_i \end{bmatrix}, \bar{D}_{fj}^q = I, \bar{C}_{ij}^q = \begin{bmatrix} -C & C \end{bmatrix}, \bar{G}_{ij}^q = \begin{bmatrix} C & 0 \end{bmatrix}, \bar{H} = \begin{bmatrix} 0 & -M \end{bmatrix}$$

For the sake of simplicity, the symbols $\mu_{j,q}(\zeta(y(t)))$ and $\Delta_{j,q}(\zeta(y(t)), \zeta(y(t_{q-1}h)))$ have been written as $\mu_{j,q}$ and $\Delta_{j,q}$, respectively.

The aim is to establish event-triggered fault detection observers for the T-S fuzzy system to detect the fault signals, and after the above analysis, the purpose is converted to solving the following problem.

Problem: For the T-S fuzzy sysyem (6.1), the event-triggered fault detection observers (6.15) are established such that the performances (S.1) and (S.2) are satisfied.

(S.1) *Asymptotic Stability:* the fault detection system (6.16) with $w(t) = 0$ is asymptotically stable.

(S.2) H_∞ Performance: the fault detection system (6.16) satisfies the following inequality:

$$\|\bar{r}(t)\|_2 \leq \beta \|w(t)\|_2 \tag{6.17}$$

where $w(t) \neq 0$, and β is the H_∞ performance index.

Moreover, the following two objectives need to be achieved: (1) The mismatch of premise variables between the system dynamics and the event-triggered fault detection observers is solved. (2) The fault signals can be successfully detected by the proposed event-triggered fault detection scheme, and better fault detection results can be obtained.

6.3 Event-Triggered Switched Fuzzy Fault Detection Observer Design

In this section, the detailed design steps of event-triggered fault detection observers will be presented by using the online asynchronous premise reconstruction approach. Next, we first give the following analysis results to provide theoretical basis for subsequent design problems.

6.3.1 Fault Detection Performance Analysis Conditions

Theorem 6.1 *For the T-S fuzzy system (6.1), design new event-triggered fault detection observers in different event intervals, given parameters σ, h, $\gamma_{1,q}$ and $\gamma_{2,q}$. The fault detection system (6.16) satisfies the performances (S.1) and (S.2), if there are* *variables* $P_s \in R^{2n\times 2n} > 0$, $P_{s+1} \in R^{2n\times 2n} > 0$, $Q \in R^{2n\times 2n} > 0$, $N \in R^{2n\times 2n} > 0$, $Z \in R^{2n\times 2n}$, $\beta > 0$, *and the fault detection observer gains* $L_{q,j}$, *such that the following inequalities hold for all* $i, j = 1, 2, \ldots, r$, $q = 1, 2, \ldots, N$, $n = 0, 1$, *and* $s = 0, 1, \ldots, L-1$.

$$\Phi^q_{iin} \le 0, \quad i = j \tag{6.18}$$

$$\Phi^q_{ijn} + \gamma_{1,q}\Phi^q_{jin} \le 0, \quad 1 \le i < j \tag{6.19}$$

$$\Phi^q_{ijn} + \gamma_{2,q}\Phi^q_{jin} \le 0, \quad 1 \le i < j \tag{6.20}$$

$$\begin{bmatrix} N & Z \\ Z^T & N \end{bmatrix} \ge 0 \tag{6.21}$$

where

$$\begin{aligned}
\Phi^q_{ijn} &= 2(\bar{A}^q_{ij}I_1 + \bar{B}^q_{ij}I_2 + \bar{E}^q_{ij}I_4 + \bar{D}^q_{ij}I_5)^T P_{s+n}I_1 + I_1^T\frac{L(P_{s+1}-P_s)}{T_q}I_1 \\
&\quad + I_1^TQI_1 - I_3^TQI_3 \\
&\quad + h^2(\bar{A}^q_{ij}I_1 + \bar{B}^q_{ij}I_2 + \bar{E}^q_{ij}I_4 + \bar{D}^q_{ij}I_5)^T N(\bar{A}^q_{ij}I_1 + \bar{B}^q_{ij}I_2 + \bar{E}^q_{ij}I_4 + \bar{D}^q_{ij}I_5) \\
&\quad - \begin{bmatrix} I_2 - I_1 \\ I_2 - I_3 \end{bmatrix}^T \begin{bmatrix} N & Z \\ Z^T & N \end{bmatrix} \begin{bmatrix} I_2 - I_1 \\ I_2 - I_3 \end{bmatrix} \\
&\quad + (\bar{C}^q_{ij}I_1 + \bar{G}^q_{ij}I_2 + \bar{D}^q_{fj}I_4 + \bar{H}I_5)^T(\bar{C}^q_{ij}I_1 + \bar{G}^q_{ij}I_2 + \bar{D}^q_{fj}I_4 + \bar{H}I_5) \\
&\quad - \beta^2 I_5^TI_5 + \sigma(\bar{H}_1I_2 + I_4)^T\Omega(\bar{H}_1I_2 + I_4) - I_4^T\Omega I_4, \\
I_1 &= \begin{bmatrix} I\,0\,0\,0\,0 \end{bmatrix}, \ldots, I_5 = \begin{bmatrix} 0\,0\,0\,0\,I \end{bmatrix}, \\
\bar{H}_1 &= \begin{bmatrix} C\;0 \end{bmatrix}
\end{aligned}$$

Proof *In this chapter, a triggering instants-dependent discontinuous Lyapunov function is constructed by using the idea of event interval partitioning. The details are given below.*

First, the event interval $[t_{q-1}h, t_qh)$ *is averagely divided into L subintervals and each subinterval is presented as* $[t_{q-1,s}h, t_{q-1,s+1}h)$, *where* $q = 1, 2, \ldots, N$, $s = 0, 1, \ldots, L-1$. *In addition, define* $T_q = LT_{q,s}$, $T_{q,s} = t_{q-1,s+1} - t_{q-1,s} = \frac{t_q - t_{q-1}}{L}$, *in which* $t_{q-1,0} = t_{q-1}$ *and* $t_{q-1,L} = t_q$. *Then, based on the above analysis, a triggering instants-dependent discontinuous Lyapunov function is constructed and presented as follows:*

$$\begin{aligned}
V(t) &= \xi^T(t)(\rho(t)P_s + \tilde{\rho}(t)P_{s+1})\xi(t) + \int_{t-h}^{t} \xi^T(s)Q\xi(s)ds \\
&\quad + h\int_{-h}^{0}\int_{t+\theta}^{t} \dot{\xi}^T(s)N\dot{\xi}(s)dsd\theta
\end{aligned} \tag{6.22}$$

in which

$$\rho(t)=\frac{t_{q-1,s+1}-t}{t_{q-1,s+1}-t_{q-1,s}},\ \tilde{\rho}(t)=1-\rho(t)$$

For the time-varying parameter $\rho(t)$, it is easy to get $\rho(t)=1$ with $t=t_{q-1,s}$ and $\rho(t)=0$ with $t=t_{q-1,s+1}$. Moreover, note that $V(t)$ is continuous in the intervals $t\in[t_{q-1,s}h, t_{q-1,s+1}h)$ and discontinuous at the triggering instants.

Next, according to (6.16), the time derivative of $V(t)$ in interval $t\in[t_{q-1,s}h, t_{q-1,s+1}h)$ is calculated and presented as follows:

$$\begin{aligned}\dot{V}(t)=&\dot{\xi}^T(t)(\rho(t)P_s+\tilde{\rho}(t)P_{s+1})\xi(t)+\xi^T(t)(\rho(t)P_s+\tilde{\rho}(t)P_{s+1})\dot{\xi}(t)\\&+\xi^T(t)(\dot{\rho}(t)P_s+\dot{\tilde{\rho}}(t)P_{s+1})\xi(t)+\xi^T(t)Q\xi(t)-\xi^T(t-h)Q\xi(t-h)\\&+h^2\dot{\xi}^T(t)N\dot{\xi}(t)-h\int_{t-h}^{t}\dot{\xi}^T(s)N\dot{\xi}(s)ds\end{aligned}\tag{6.23}$$

It should be pointed out that $\rho(t)$ is differentiable in $[t_{q-1,s}h, t_{q-1,s+1}h)$ and is non-differentiable for all $t\geq 0$. Therefore, the Dini derivative is introduced to cope with it, and then the following equation is established:

$$\begin{aligned}D^+(\rho(t)P_s+\tilde{\rho}(t)P_{s+1})&=\frac{-1}{t_{q-1,s+1}-t_{q-1,s}}P_s+\frac{1}{t_{q-1,s+1}-t_{q-1,s}}P_{s+1}\\&=\frac{L(P_{s+1}-P_s)}{T_q}\end{aligned}\tag{6.24}$$

Furthermore, define new variable:

$$J(t)=D^+V(t)+\bar{r}^T(t)\bar{r}(t)-\beta^2w^T(t)w(t)\tag{6.25}$$

Then, one obtains

$$\begin{aligned}J(t)=&D^+V(t)+\bar{r}^T(t)\bar{r}(t)-\beta^2w^T(t)w(t)\\\leq&\dot{\xi}^T(t)(\rho(t)P_s+\tilde{\rho}(t)P_{s+1})\xi(t)+\xi^T(t)(\rho(t)P_s+\tilde{\rho}(t)P_{s+1})\dot{\xi}(t)\\&+\xi^T(t)\left(\frac{L(P_{s+1}-P_s)}{T_q}\right)\xi(t)+\xi^T(t)Q\xi(t)-\xi^T(t-h)Q\xi(t-h)\\&+h^2\dot{\xi}^T(t)N\dot{\xi}(t)-h\int_{t-h}^{t}\dot{\xi}^T(s)N\dot{\xi}(s)ds+\bar{r}^T(t)\bar{r}(t)-\beta^2w^T(t)w(t)\end{aligned}\tag{6.26}$$

Let $\begin{bmatrix}N & Z\\ Z^T & N\end{bmatrix}\geq 0$, and then use the reciprocally convex method to obtain:

$$-h\int_{t-h}^{t}\dot{\xi}^T(s)N\dot{\xi}(s)ds$$
$$\leq -\begin{bmatrix}\xi(t-\tau(t))-\xi(t)\\ \xi(t-\tau(t))-\xi(t-h)\end{bmatrix}^T\begin{bmatrix}N & Z\\ Z^T & N\end{bmatrix}\begin{bmatrix}\xi(t-\tau(t))-\xi(t)\\ \xi(t-\tau(t))-\xi(t-h)\end{bmatrix} \tag{6.27}$$

Thus, one can further construct the following inequality:

$$\begin{aligned}J(t) &= D^+V(t)+\bar{r}^T(t)\bar{r}(t)-\beta^2 w^T(t)w(t)\\ &\leq \dot{\xi}^T(t)(\rho(t)P_s+\tilde{\rho}(t)P_{s+1})\xi(t)+\xi^T(t)(\rho(t)P_s+\tilde{\rho}(t)P_{s+1})\dot{\xi}(t)\\ &\quad+\xi^T(t)\left(\frac{L(P_{s+1}-P_s)}{T_q}\right)\xi(t)+\xi^T(t)Q\xi(t)-\xi^T(t-h)Q\xi(t-h)\\ &\quad+h^2\dot{\xi}^T(t)N\dot{\xi}(t)+\bar{r}^T(t)\bar{r}(t)-\beta^2 w^T(t)w(t)\\ &\quad-\begin{bmatrix}\xi(t-\tau(t))-\xi(t)\\ \xi(t-\tau(t))-\xi(t-h)\end{bmatrix}^T\begin{bmatrix}N & Z\\ Z^T & N\end{bmatrix}\begin{bmatrix}\xi(t-\tau(t))-\xi(t)\\ \xi(t-\tau(t))-\xi(t-h)\end{bmatrix}\\ &\quad+\sigma(y(t-\tau(t))+e(t))^T\Omega(y(t-\tau(t))+e(t))-e^T(t)\Omega e(t)\end{aligned} \tag{6.28}$$

Moreover, introduce new variable $\epsilon(t)=[\xi^T(t),\xi^T(t-\tau(t)),\xi^T(t-h),e^T(t),w^T(t)]^T$*, and then*

$$\begin{aligned}J(t) &= D^+V(t)+\bar{r}^T(t)\bar{r}(t)-\beta^2 w^T(t)w(t)\\ &\leq \dot{\xi}^T(t)P_{s+n}\xi(t)+\xi^T(t)P_{s+n}\dot{\xi}(t)+\xi^T(t)Q\xi(t)\\ &\quad+\xi^T(t)\left(\frac{L(P_{s+1}-P_s)}{T_q}\right)\xi(t)-\xi^T(t-h)Q\xi(t-h)\\ &\quad+h^2\dot{\xi}^T(t)N\dot{\xi}(t)+\bar{r}^T(t)\bar{r}(t)-\beta^2 w^T(t)w(t)\\ &\quad-\begin{bmatrix}\xi(t-\tau(t))-\xi(t)\\ \xi(t-\tau(t))-\xi(t-h)\end{bmatrix}^T\begin{bmatrix}N & Z\\ Z^T & N\end{bmatrix}\begin{bmatrix}\xi(t-\tau(t))-\xi(t)\\ \xi(t-\tau(t))-\xi(t-h)\end{bmatrix}\\ &\quad+\sigma(y(t-\tau(t))+e(t))^T\Omega(y(t-\tau(t))+e(t))-e^T(t)\Omega e(t)\\ &= \sum_{i=1}^{r}\sum_{j=1}^{r}\Delta_{j,q}\mu_i\mu_j\epsilon^T(t)\Phi^q_{ijn}\epsilon(t)\\ &= \sum_{i=1}^{r-1}\sum_{j>i}^{r}\Delta_{j,q}\mu_i\mu_j(\Phi^q_{ijn}+\frac{\Delta_{i,q}}{\Delta_{j,q}}\Phi^q_{jin})+\sum_{i=1}^{r}\Delta_{i,q}\mu_i^2\epsilon^T(t)\Phi^q_{iin}\epsilon(t)\end{aligned} \tag{6.29}$$

in which Φ^q_{ijn} *has been described in Theorem 6.1.*

By applying the convex combination technique to (6.29), it is known that the inequalities (6.18)–(6.21) ensure $J(t)<0$ *holds, which implies that the fault detection system (6.16) satisfies the* H_∞ *performance. In addition, when* $w(t)=0$*, it is also* *concluded from (6.18)–(6.21) that the fault detection system (6.16) is asymptotically stable.*

The performance analysis results have been summarized as Theorem 6.1, in which it is important to note that a triggering instants-dependent discontinuous Lyapunov function is established to guarantee the performances of the fault detection system. When $\rho(t) = 1$ is selected, the analysis results established in this chapter will be reduced to the following corollary.

Corollary 6.1 *For the T-S fuzzy system (6.1), design new event-triggered fault detection observers in different event intervals, given parameters* σ, h, $\gamma_{1,q}$ *and* $\gamma_{2,q}$. *The fault detection system (6.16) satisfies the performances (S.1) and (S.2), if there are variables* $Q \in R^{2n\times 2n} > 0$, $N \in R^{2n\times 2n} > 0$, $Z \in R^{2n\times 2n}$, $\beta > 0$, *and the fault detection observer gains* $L_{q,j}$, *such that (6.30)–(6.33) hold for all* $i, j = 1, 2, \ldots, r$, $q = 1, 2, \ldots, N$.

$$\Re_{ii}^q \le 0, \quad i = j \tag{6.30}$$

$$\Re_{ij}^q + \gamma_{1,q}\Re_{ji}^q \le 0, \quad 1 \le i < j \tag{6.31}$$

$$\Re_{ij}^q + \gamma_{2,q}\Re_{ji}^q \le 0, \quad 1 \le i < j \tag{6.32}$$

$$\begin{bmatrix} N & Z \\ Z^T & N \end{bmatrix} \ge 0 \tag{6.33}$$

where

$$\begin{aligned}
\Re_{ij}^q &= 2(\bar{A}_{ij}^q I_1 + \bar{B}_{ij}^q I_2 + \bar{E}_{ij}^q I_4 + \bar{D}_{ij}^q I_5)^T P I_1 + I_1^T Q I_1 - I_3^T Q I_3 \\
&\quad + h^2(\bar{A}_{ij}^q I_1 + \bar{B}_{ij}^q I_2 + \bar{E}_{ij}^q I_4 + \bar{D}_{ij}^q I_5)^T N (\bar{A}_{ij}^q I_1 + \bar{B}_{ij}^q I_2 + \bar{E}_{ij}^q I_4 + \bar{D}_{ij}^q I_5) \\
&\quad - \begin{bmatrix} I_2 - I_1 \\ I_2 - I_3 \end{bmatrix}^T \begin{bmatrix} N & Z \\ Z^T & N \end{bmatrix} \begin{bmatrix} I_2 - I_1 \\ I_2 - I_3 \end{bmatrix} \\
&\quad + (\bar{C}_{ij}^q I_1 + \bar{G}_{ij}^q I_2 + \bar{D}_{fj}^q I_4 + \bar{H} I_5)^T (\bar{C}_{ij}^q I_1 + \bar{G}_{ij}^q I_2 + \bar{D}_{fj}^q I_4 + \bar{H} I_5) \\
&\quad - \beta^2 I_5^T I_5 + \sigma(\bar{H}_1 I_2 + I_4)^T \Omega (\bar{H}_1 I_2 + I_4) - I_4^T \Omega I_4
\end{aligned}$$

Proof The Lyapunov function is defined as follows:

$$V(t) = \xi^T(t) P \xi(t) + \int_{t-h}^{t} \xi^T(s) Q \xi(s) ds + h \int_{-h}^{0} \int_{t+\theta}^{t} \dot{\xi}^T(s) N \dot{\xi}(s) ds d\theta \tag{6.34}$$

The proof is similar to that given in Refs. [2, 11], so omitted here.

Remark 6.2 Note that the triggering instants are not effectively utilized in the existing event-triggered results, and therefore, in order to achieve better fault detection performances, a triggering instants-dependent discontinuous Lyapunov function is constructed by using the idea of event interval partitioning, in which the event interval is averagely divided into L subintervals, and a larger L can help to achieve better fault detection results but may increase computational complexity. Therefore, a trade-off is necessary.

Remark 6.3 Theorem 6.1 and Corollary 6.1 respectively summarize the fault detection results proposed in this chapter and the existing results in Refs. [2, 11] using the constant Lyapunov functions and the proposed asynchronous premise reconstruction method. When $\rho(t) = 1$ is selected, the analysis results established in this paper will be reduced to Corollary 6.1.

6.3.2 Event-Triggered Switched Fuzzy Fault Detection Observer Design Conditions

Based on the analysis results established above, the event-triggered fault detection observer design results are further presented and summarized as Theorem 6.2.

Theorem 6.2 *For the T-S fuzzy system (6.1), design new event-triggered fault detection observers in different event intervals, given parameters* σ, h, $\gamma_{1,q}$ *and* $\gamma_{2,q}$. *The fault detection system (6.16) satisfies the performances (S.1) and (S.2), if there are variables* $\begin{bmatrix} Y_s & -R_s \\ -R_s & R_s \end{bmatrix} > 0$, $\begin{bmatrix} Y_{s+1} & -R_{s+1} \\ -R_{s+1} & R_{s+1} \end{bmatrix} > 0$, $Q \in R^{2n\times 2n} > 0$, $N \in R^{2n\times 2n} > 0$, $Z \in R^{2n\times 2n}$, $\beta > 0$, *and the matrix variables* $W_{j,q,s} \in R^{n\times m}$ *and* $W_{j,q,s+1} \in R^{n\times m}$, *such that (6.35)–(6.38) hold for all* $i, j = 1, 2, \ldots, r$, $q = 1, 2, \ldots, N$, $n = 0, 1$, *and* $s = 0, 1, \ldots, L-1$.

$$\Phi_{iin}^{q} \le 0, \quad i = j \tag{6.35}$$

$$\Phi_{ijn}^{q} + \gamma_{1,q}\Phi_{jin}^{q} \le 0, \quad 1 \le i < j \tag{6.36}$$

$$\Phi_{ijn}^{q} + \gamma_{2,q}\Phi_{jin}^{q} \le 0, \quad 1 \le i < j \tag{6.37}$$

$$\begin{bmatrix} N & Z \\ Z^T & N \end{bmatrix} \ge 0 \tag{6.38}$$

Then, the gains of the event-triggered fault detection observers are calculated by the following equation:

$$L_{j,q} = R_s^{-1} W_{j,q,s}$$

Proof First, the matrix variable P_s is selected as $P_s = \begin{bmatrix} Y_s & -R_s \\ -R_s & R_s \end{bmatrix}$, and define $R_s L_{j,q}$ as a new variable $W_{j,q,s}$. Then, the inequalities (6.18)–(6.21) are linearized.

6.4 On-Line Algorithm

Based on the designed online asynchronous premise reconstruction method, new fault detection observer design results are constructed and summarized as Theorem 6.2. Next, in order to calculate the gains of the event-triggered fault detection observers, an online algorithm is designed and described as follows.

Event-triggered fault detection observers design

Step 1: $q = 1$.

(1) Set the initial data to be transmitted such that $\mu_j(\zeta(y(t_0 h)))$ is obtained.

(2) $\gamma_{1,1} = 1$ and $\gamma_{2,1} = 1$ are defined from (6.10)–(6.14).

(3) Solve the inequalities (6.35)–(6.38) with $q = 1, n = 0, 1, s = 0, 1, \ldots, L-1$, $\gamma_{1,1} = 1$ and $\gamma_{2,1} = 1$ to get the initial gains $L_{1,1}, \ldots, L_{r,1}$.

(4) Go to Step 2.

Step 2: $q = 2$.

(1) Obtain the current membership function $\mu_j(\zeta(y(t)))$ in the event-triggered unit in the interval $[t_0 h, t_1 h)$ by executing step 1.

(2) Compute the parameters $\gamma_{1,2}$ and $\gamma_{2,2}$ online in the interval $[t_0 h, t_1 h)$ by the inequalities (6.10)–(6.14) and $\mu_j(\zeta(y(t_0 h)))$.

(3) Solve the inequalities (6.35)–(6.38) with $q = 2, n = 0, 1, s = 0, 1, \ldots, L-1$, $\gamma_{1,2}$ and $\gamma_{2,2}$ to obtain the gains $L_{1,2}, \ldots, L_{r,2}$.

(4) Go to Step 3.

$\vdots$

Step N: $q = N$.

(1) Obtain the current membership function $\mu_j(\zeta(y(t)))$ in the event-triggered unit in the interval $[t_{N-2}h, t_{N-1}h)$ by executing step $N-1$.

(2) Compute the parameters $\gamma_{1,N}$ and $\gamma_{2,N}$ online in the interval $[t_{N-2}h, t_{N-1}h)$ by the inequalities (6.10)–(6.14) and $\mu_j(\zeta(y(t_{N-2}h)))$.

(3) Solve the inequalities (6.35)–(6.38) with $q = N, n = 0, 1, s = 0, 1, \ldots, L-1$, $\gamma_{1,N}$ and $\gamma_{2,N}$ to obtain the gains $L_{1,N}, \ldots, L_{r,N}$.

(4) Stop running.

Remark 6.4 It should be explained that in interval $[t_{q-1}h, t_q h)$, the membership function $\mu_j(\zeta(y(t)))$ is completely known in the event-triggered unit, and the minimum and maximum values of $\Delta_{i,k}$ are solved by (6.10)–(6.14) and $\mu_j(\zeta(y(t_{q-1}h)))$, and obtained online in the event-triggered unit. When the next data is transmitted by the event unit, some information, including $\gamma_{1,q+1}$, $\gamma_{2,q+1}$, $t_q h$ and $y(t_q h)$, will be packaged and transmitted to the event-triggered fault detection observers. Moreover, note that when the membership functions $\mu_i(\zeta(y(t))) = 0$ or $\mu_i(\zeta(y(t_{q-1}h))) = 0$, the corresponding upper and lower bounds of γ will be assumed to be $\gamma_{1,q} = 1$ and $\gamma_{2,q} = 1$.

6.5 Numeral Example

In this section, the effectiveness and superiority of the event-triggered fault detection method designed in this chapter will be verified through the tunnel diode circuit in Ref. [12], in which the tunnel diode circuit is plotted in Figure 2.1 of Chap. 2, and the model is described as follows:

$$i_D(t) = 0.002v_D(t) + 0.01v_D^3(t)$$

Define the state variables as $x_1(t) = v_C(t)$ and $x_2(t) = i_L(t)$, Then, the model can be converted to the following state equation:

$$\begin{aligned} C\dot{x}_1(t) &= -0.002x_1(t) - 0.01x_1^3(t) + x_2(t) \\ L\dot{x}_2(t) &= -x_1(t) - Rx_2(t) + v(t) \\ y(t) &= x_1(t) \end{aligned} \tag{6.39}$$

where the parameter matrices are $C = 20\,\text{mF}$, $L = 1000\,\text{mH}$ and $R = 10\,\Omega$, respectively, and the controller $u(t) = Ky(t)$ has been designed to stable the system model.

It is assumed that there are bias faults in the capacitor voltage, and then similar to the Ref. [12], the state equation described above can be established as the T-S fuzzy model (6.1), in which the system matrices and the fault matrices are shown as follows:

$$\begin{aligned} A_1 &= \begin{bmatrix} -0.1 & 50 \\ -1 & -10 \end{bmatrix}, A_2 = \begin{bmatrix} -4.6 & 50 \\ -1 & -10 \end{bmatrix}, D_1 = \begin{bmatrix} 0 \\ 1 \end{bmatrix}, \\ D_2 &= \begin{bmatrix} 0 \\ 1 \end{bmatrix}, G_1 = \begin{bmatrix} 0 \\ 1 \end{bmatrix}, G_2 = \begin{bmatrix} 1 \\ 0.5 \end{bmatrix}, C = \begin{bmatrix} 1 & 0 \end{bmatrix} \end{aligned}$$

The membership functions of the T-S fuzzy system are chosen to be the same as in Ref. [12], and are shown in Fig. 6.1, where the premise variables depend on the system outputs $y(t)$. Moreover, note that the measurement outputs may not be accurately received by the fault detection observers due to the existence of event-triggered mechanism, and the outputs $y(t_{q-1}h)$ are transmitted to the fault detection observer side and used to establish the event-triggered fault detection observers. Therefore, $y(t_{q-1}h)$ is selected as the premise variables of fault detection observers, and the membership functions of the fuzzy fault detection observers are the same as those of the T-S system model.

The purpose is to design event-triggered fault detection observers for the T-S fuzzy system to detect faults accurately. In order to reduce the conservatism of the existing method in Ref. [2], the idea of event interval partitioning is exploited, and then new fault detection results are presented. In this example, the event interval $[t_{q-1}h, t_qh)$ is averagely divided into L subintervals, and $L = 3$, 2 and 1 are, respectively, chosen to illustrate the effectiveness of the designed fault detection scheme.

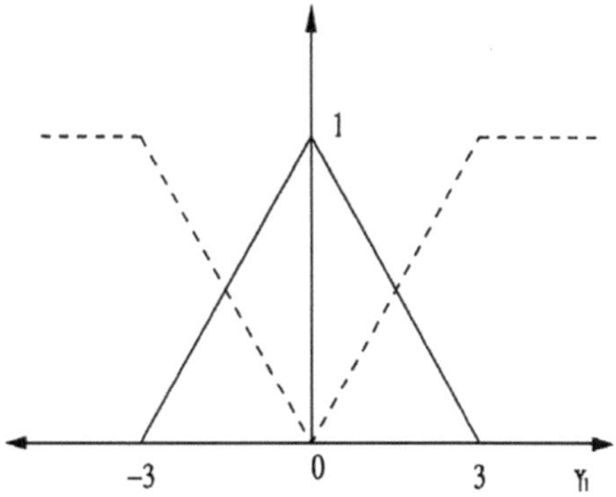

Fig. 6.1 Membership functions

In addition, the existing result in Ref. [2] using the constant Lyapunov function and the proposed asynchronous premise reconstruction method have also been considered in this example.

For the four cases, the parameters $\gamma_{1,1} = 1$ and $\gamma_{2,1} = 1$ are first chosen because the initial data are transmitted. Then, the initial parameter matrices are calculated respectively by solving the inequalities (6.35)–(6.38) with $\gamma_{1,1} = 1$, $\gamma_{2,1} = 1$, $\sigma = 0.06$, the inequalities (6.30)–(6.33), and presented as follows:

Case 1: Theorem 6.2 with $L = 3$.

$$L_{1,1} = \begin{bmatrix} 0.3693 & -0.0405 \end{bmatrix}, L_{2,1} = \begin{bmatrix} -1.0833 & 0.1738 \end{bmatrix}, \Omega_1 = 159.3000$$

Case 2: Theorem 6.2 with $L = 2$.

$$L_{1,1} = \begin{bmatrix} 0.4461 & -0.0570 \end{bmatrix}, L_{2,1} = \begin{bmatrix} -1.0107 & 0.1560 \end{bmatrix}, \Omega_1 = 204.7485$$

Case 3: Theorem 6.2 with $L = 1$.

$$L_{1,1} = \begin{bmatrix} 0.6137 & -0.0912 \end{bmatrix}, L_{2,1} = \begin{bmatrix} -0.8053 & 0.1100 \end{bmatrix}, \Omega_1 = 275.3872$$

Case 4: Corollary 6.1 (the existing results in Refs. [2, 11]).

$$L_{1,1} = \begin{bmatrix} 1.5806 & -0.2806 \end{bmatrix}, L_{2,1} = \begin{bmatrix} 0.7152 & -0.1740 \end{bmatrix}, \Omega_1 = 30.9382$$

In order to better demonstrate the effectiveness of the presented fault detection scheme, the periodic disturbance $v(t) = 0.5sin(0.2t)$ is considered and enter the system at $t = 0$, and the periodic fault signal $f(t) = 1.5 + 1.5sin(0.1t)$ will be generated at $t = 15$ in the actuators.

First, based on the initial parameter matrices computed above, the main results are obtained by implementing the proposed online algorithm and presented in Figs. 6.2, 6.3, 6.4 and 6.5, in which the residual evaluation outputs with $L = 3$, $L = 2$, $L = 1$ are plotted in Figure 6.2. Figure 6.2 shows that the faults are, respectively, detected by

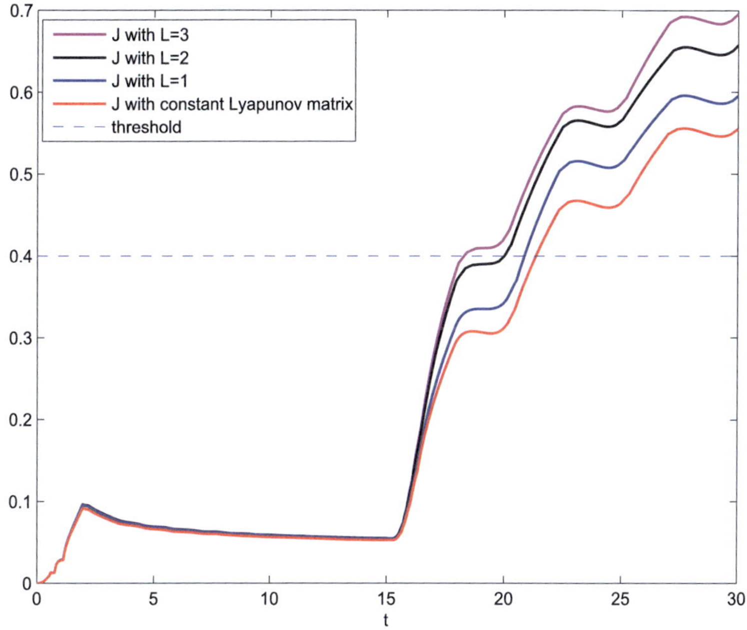

Fig. 6.2 Residual evaluation outputs with $L = 3$, $L = 2$, $L = 1$ and constant Lyapunov matrix

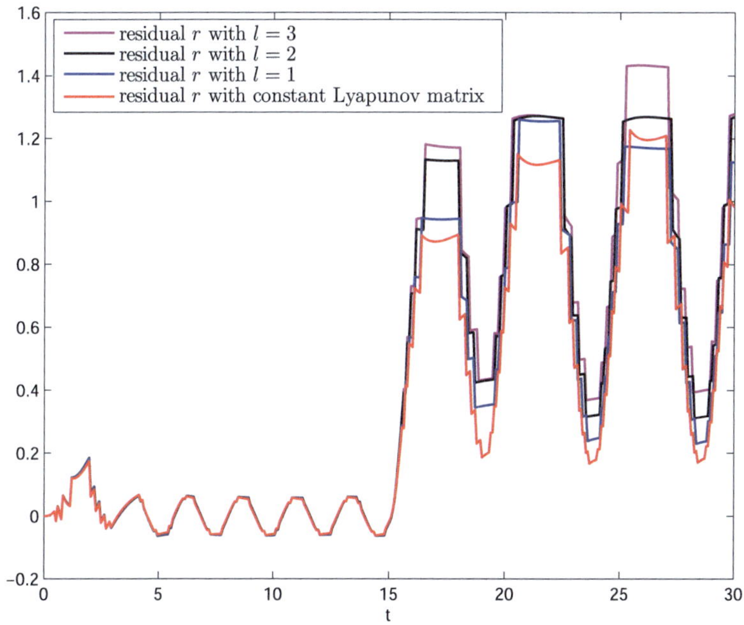

Fig. 6.3 Residual signals with $L = 3$, $L = 2$, $L = 1$ and constant Lyapunov matrix

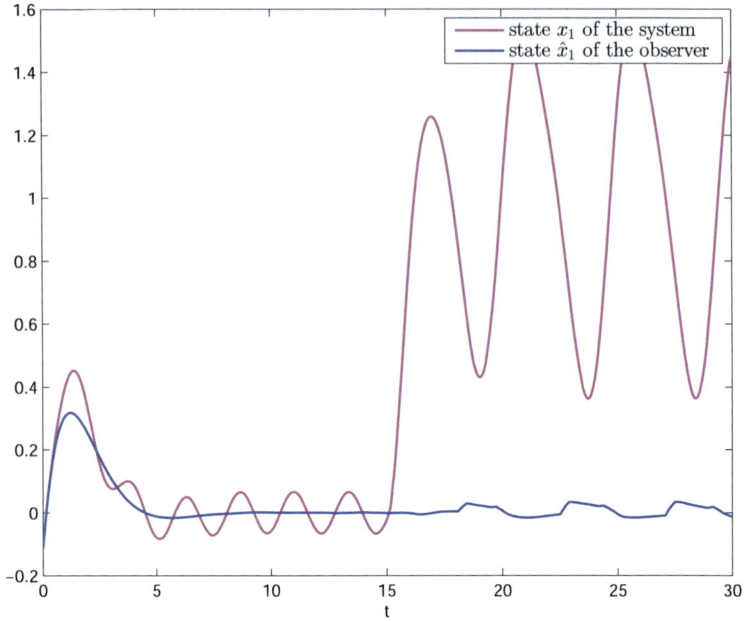

Fig. 6.4 States x_1 and $\hat{x}_1$

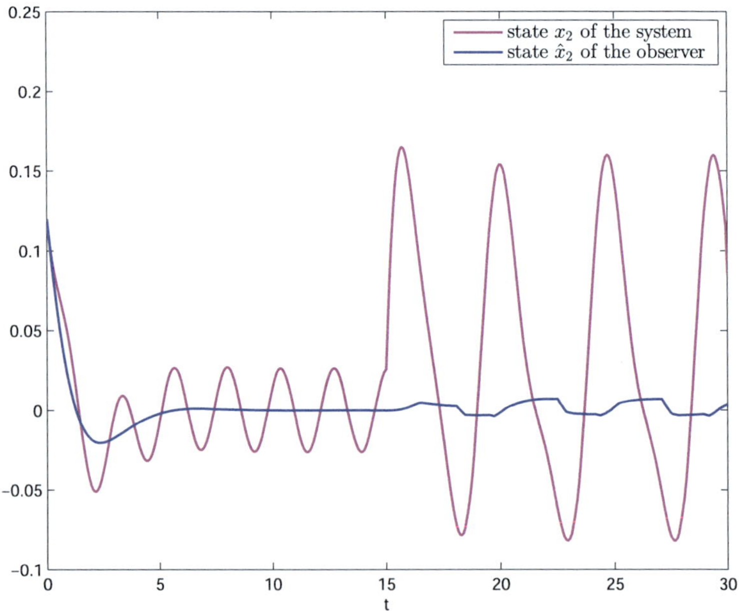

Fig. 6.5 States x_2 and $\hat{x}_2$

applying the designed method with $L = 3, L = 2, L = 1$ at approximately $k = 18.3$, $k = 20.0$ and $k = 20.8$, which indicates that more event subintervals are partitioned, better fault detection performances are obtained, and at the same time, more computation is needed.

Then, in order to demonstrate the superiority of the established fault detection scheme, the residual evaluation outputs of the existing method in Ref. [2] using the constant Lyapunov functions and the proposed asynchronous premise reconstruction scheme are also plotted in Fig. 6.2, respectively. It is seen that the fault is detected at approximately $k = 21.3$. Thus, it is concluded that the designed approach achieves better fault detection results than that in Ref. [2]. The residual signals with $L = 3$, $L = 2$, $L = 1$ and constant Lyapunov function are presented in Fig. 6.3. From Fig. 6.3, it is indicated that the disturbance in T-S fuzzy system is suppressed in the absence of fault, and the residual signal will jump when the fault occurs. The states of the T-S fuzzy system and fault detection observers are presented in Figs. 6.4 and 6.5.

In addition, in order to further illustrate the effectiveness of the method designed in this chapter, case 1 is selected for detailed explanation, and the main results are presented in Figs. 6.6, 6.7, 6.8 and 6.9, in which Figs. 6.6 and 6.7 present the gains of the event-triggered fault detection observers, the release instants and release intervals of the event-triggered mechanism are plotted in Fig. 6.8, and from which one can further know that the gains of the event-triggered fault detection observers will

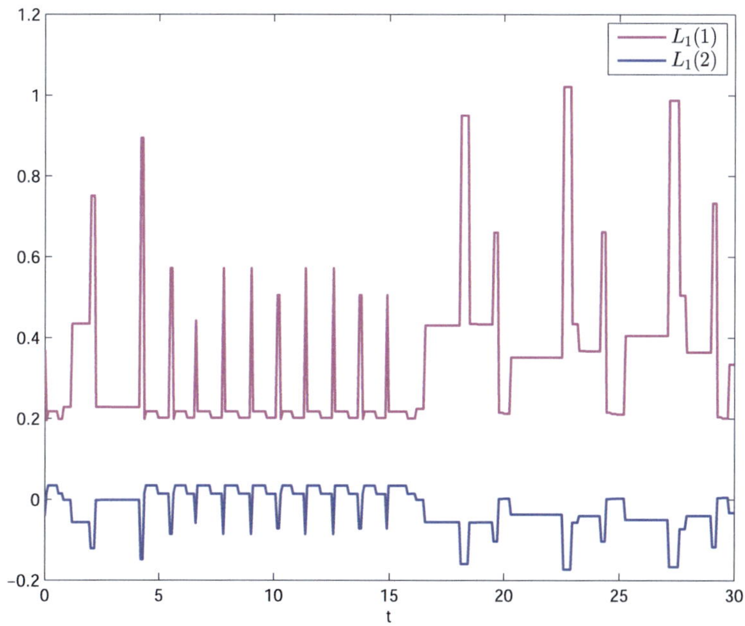

Fig. 6.6 Fault detection observer gains $L_{1,q}$

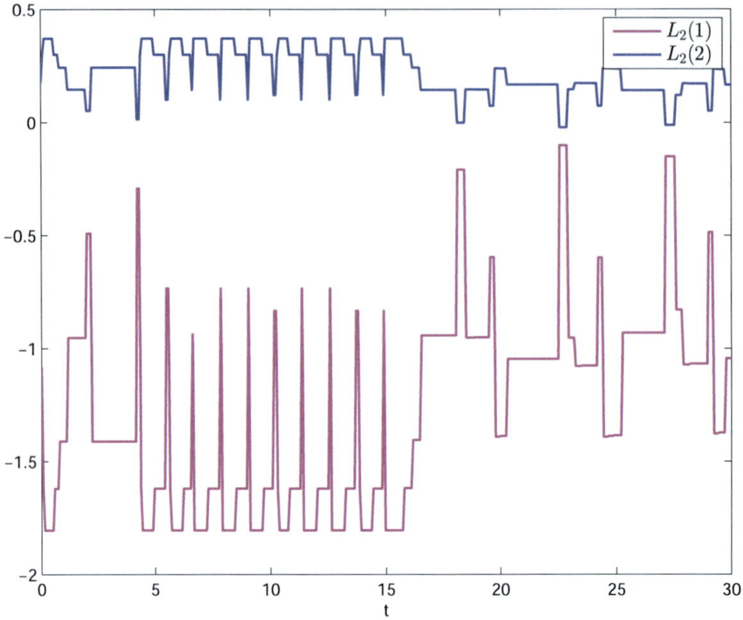

Fig. 6.7 Fault detection observer gains $L_{2,q}$

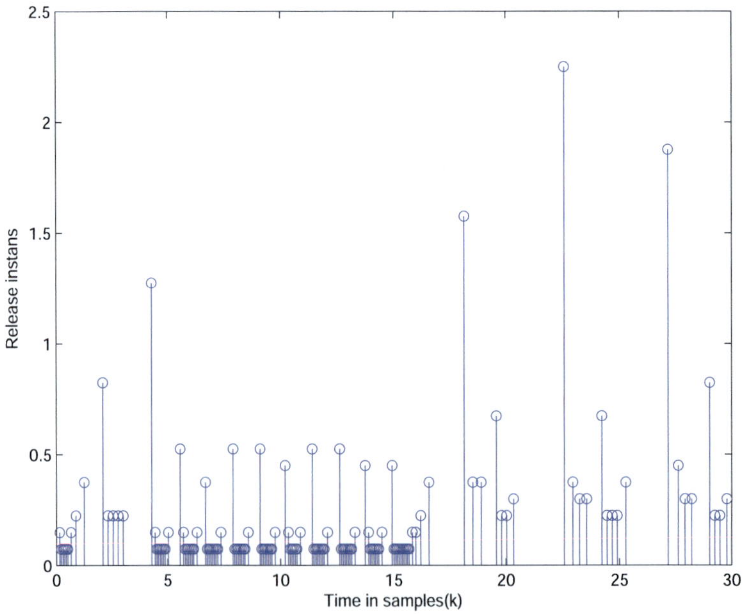

Fig. 6.8 Release instants and time intervals

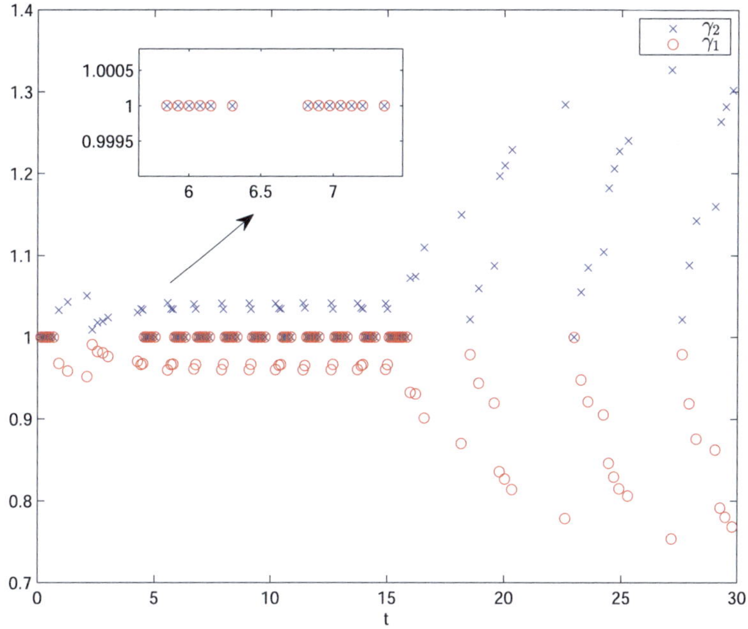

Fig. 6.9 γ_1 and γ_2

be updated at different triggering instants and maintained until the next triggering instant. Moreover, the parameters $\gamma_{1,q}$ and $\gamma_{2,q}$ are obtained by implementing the designed algorithm and plotted in Fig. 6.9, where it should be explained that when the membership functions $\mu_i(\zeta(y(t))) = 0$ or $\mu_i(\zeta(y(t_{q-1}h))) = 0$, the corresponding upper and lower bounds of γ will be assumed to be $\gamma_{1,q} = 1$ and $\gamma_{2,q} = 1$.

6.6 Summary

This chapter has investigated the problem of designing event-triggered fault detection observers for T-S fuzzy systems. Assume that the system outputs are selected as the premise variables of T-S fuzzy systems. First, online asynchronous premise reconstruction method is proposed to address the challenges incurred by the mismatch of premise variables. Then, new types of observers, namely, switched fuzzy fault detection observers, are established, which are switched at different triggering instants to detect the faults. Subsequently, triggering instants-dependent discontinuous Lyapunov functions are constructed by using the idea of event interval partitioning which is exploited to achieve better fault detection results. Finally, an example is presented to verify the validity of the proposed design technique.

6.7 Notes

The methods proposed in Refs. [7, 8] cannot be effectively used to solve the asynchronous problem of premise variables, and it is necessary to construct a new asynchronous premise reconstruction method to improve the defects in these methods. Moreover, better fault detection results will be facilitated by making full use of the triggering instant information. Therefore, how to effectively utilize the triggering instants to establish a better fault detection scheme is an important and meaningful issue, which motivates this paper. In addition, it should be noted that the results presented in this chapter are mainly constructed based on the ones in Ref. [13].

References

1. Wang XL, Yang GH (2020) Distributed event-triggered H_∞ filtering for discrete-time T-S fuzzy systems over sensor networks. IEEE Trans Syst Man Cybern: Syst 50(9):3269–3280
2. Zhang CZ, Hu JF, Qiu JB, Chen QJ (2017) Event-triggered nonsynchronized H_∞ filtering for discrete-time T-S fuzzy systems based on piecewise Lyapunov functions. IEEE Trans Syst Man Cybern: Syst 47(8):2330–2341
3. Lam HK, Narimani M (2009) Stability analysis and performance design for fuzzy-model-based control system under imperfect premise matching. IEEE Trans Fuzzy Syst 17(4):949–961
4. Peng C, Ma SD, Xie XP (2017) Observer-based non-PDC control for networked T-S fuzzy systems with an event-triggered communication. IEEE Trans Cybern 47(8):2279–2287
5. Jin YS, Kwon WY, Lee S (2020) Further results on sampled-data H_∞ filtering for T-S fuzzy systems with asynchronous premise variables. IEEE Trans Fuzzy Syst 30(6):1864–1874
6. Lee S (2019) Novel stabilization criteria for T-S fuzzy systems with affine matched membership functions. IEEE Trans Fuzzy Syst 27(3):540–548
7. Peng C, Yang MJ, Zhang J, Fei MR, Hu SL (2017) Network-based H_∞ control for T-S fuzzy systems with an adaptive event-triggered communication scheme. Fuzzy Sets Syst 329:61–76
8. Liu YJ, Guo BZ, Park JH, Lee S (2018) Event-based reliable dissipative filtering for T-S fuzzy systems with asynchronous constraints. IEEE Trans Fuzzy Syst 26(4):2089–2098
9. Liu K, Fridman E, Hetel L (2012) Stability and L_2-gain analysis of networked control systems under round-robin scheduling: a time-delay approach. Syst Control Lett 61(5):666–675
10. Liu K, Fridman E (2012) Networked-based stabilization via discontinuous Lyapunov functionals. Int J Robust Nonlinear Control 22(4):420–436
11. Wang XL, Yang GH (2019) Event-triggered H_∞ filtering for discrete-time T-S fuzzy systems via network delay optimization technique. IEEE Trans Syst Man Cybern: Syst 49(10):2026–2035
12. Nguang SK, Assawinchaichote W (2003) H_∞ filtering for fuzzy dynamical systems with D stability constraints. IEEE Trans Circ Syst I: Fundam Theory Appl 50(11):1503–1508
13. Wang XL, Yang GH, Zhang DH (2021) Event-triggered fault detection observer design for T-S fuzzy systems. IEEE Trans Fuzzy Syst 29(9):2532–2542

Chapter 7
Fuzzy Fault Detection Observer Design for T-S Fuzzy Systems With Measurement Outliers

7.1 Introduction

The T-S fuzzy model is considered to be able to approximate nonlinear systems due to its precise approximation ability [1–4], which prompts the study of T-S fuzzy systems in this book. The filtering problems of T-S fuzzy systems have been studied in Chaps. 3–5. In Chap. 6 and this chapter, the design of event-triggered fault detection observers for T-S fuzzy systems has also been discussed, and the corresponding fault detection methods in Refs. [5–11] were introduced. For example, a switching mechanism in Ref. [8] was introduced for designing fault detection filters for T-S fuzzy systems with unknown membership functions. In Refs. [9, 10], the fault detection filters were constructed to detect faults for discrete-time affine T-S fuzzy systems and T-S fuzzy systems, respectively. In Ref. [11], the problem of fault detection for interval type-2 fuzzy systems with event-triggered mechanism was investigated, in which the asynchronous problem of the premise variables caused by the introduction of event-triggered mechanism is addressed. Note that the aforementioned methods can effectively deal with the problems of fault detection. However, it is worth mentioning that the fault detection alarm devices are usually affected by measurement outliers, such as pulse disturbances that may completely destroy the output measurements used to build the residual signals, leading to the generation of false alarms. Therefore, it is necessary to reduce the impact of measurement outliers on the systems.

Many achievements have been made in the study of measurement outliers in the past few decades, such as in Refs. [12, 13], the main idea of dealing with outliers is that the established Kalman filter is robust to measurement outliers. In Ref. [14], a statistical test method for the insensitivity to measurement outliers was designed to handle the possible outliers. The state estimation problem of linear systems with measurement outliers was studied in Ref. [15], where a more effective method based on a moving-horizon strategy is proposed. In Ref. [16], the problem of state estimation in the presence of outliers in the system was studied, in which a new observer with

X.-L. Wang et al., *Robust Filtering and Fault Detection for T-S Fuzzy Systems*,
https://doi.org/10.1007/978-981-96-5818-3_7

a saturated output injection is constructed to attenuate the impact of measurement outliers, and the stability conditions of the error system are further derived. However, it is important to note that the existing results mainly focus on the problems of state estimation for linear systems, and how to establish fuzzy fault detection observer for T-S fuzzy systems to reduce the influence of measurement outliers on the estimation errors and residual signals, so as to avoid the occurrence of false alarms, is a problem that has not been investigated. Therefore, it is of great theoretical significance and practical value to research this problem.

This paper investigates the fault detection problem for T-S fuzzy systems subject to measurement outliers. The adaptive saturation of output errors is used to construct a new fuzzy fault detection observer, such that the measurement outliers cannot be directly applied to the error systems and the residual signals, thereby avoiding the occurrence of false alarms. By establishing a suitable non-differentiable Lyapunov function, exponential stability conditions and H_∞ performance conditions of the error systems are presented in terms of convex problem constrained by linear matrix inequalities. Moreover, the proposed fault detection method with adaptive saturation of output errors can be reduced to the existing methods with typical output errors. Finally, simulation examples are provided to verify the effectiveness of the proposed design strategy.

7.2 System Modeling and Problem Statement

A class of continuous-time T-S fuzzy systems is considered, as shown below:

$$\begin{aligned}\dot{x}(t) &= \sum_{i=1}^{r} h_i(z(t))(A_i x(t) + D_i v(t) + F_i f(t))\\ &= A_h x(t) + D_h v(t) + F_h f(t)\\ y(t) &= Cx(t)\end{aligned} \tag{7.1}$$

with

$$A(h) = \sum_{i=1}^{r} h_i(z(t))A_i,\ D_h = \sum_{i=1}^{r} h_i(z(t))D_i,\ F_h = \sum_{i=1}^{r} h_i(z(t))F_i$$

where $i = 1, 2, \ldots, r$, and r is the number of **IF-THEN** rules of the T-S fuzzy system. $x(t) \in R^n$ is the state vector. $z(t)$ represents the premise variables that depend on the system states. $v(t) \in R^v$ is the disturbance belonging to $L_2[0, \infty)$. $f(t) \in R^u$ is the fault signal. $y(t) \in R^m$ is the output vector. A_i, D_i, F_i, C are known constant matrices. Moreover, the membership functions $h_i(z(t))$ $(i = 1, 2, \ldots, r)$ satisfy the following conditions:

$$h_i(z(t)) \geq 0, \sum_{i=1}^{r} h_i(z(t)) = 1, i = 1, 2, \ldots, r \tag{7.2}$$

The outliers may interfere with the outputs of the system, deteriorate the performance of the system and even make the system unstable. Therefore, in order to effectively deal with the impact of measurement outliers on the systems, a new fuzzy fault detection observer is constructed in this chapter for the T-S fuzzy system, which is presented as follows:

$$\begin{aligned}\dot{\hat{x}}(t) &= \sum_{j=1}^{r} h_j(z(t))(A_j\hat{x}(t) + L_j sat_{\sigma(t)}(y(t) - \hat{y}(t))) \\ &= A_{\hat{h}}\hat{x}(t) + L_{\hat{h}} sat_{\sigma(t)}(y(t) - \hat{y}(t)) \\ \hat{y}(t) &= C\hat{x}(t) \\ r(t) &= sat_{\sigma(t)}(y(t) - \hat{y}(t))\end{aligned} \tag{7.3}$$

with

$$A_{\hat{h}} = \sum_{j=1}^{r} h_j(z(t))A_j, L_{\hat{h}} = \sum_{j=1}^{r} h_j(z(t))L_j$$

in which $\hat{x}(t) \in R^n$ and $\hat{y}(t) \in R^m$ are the estimation of the state and the output, respectively. $r(t) \in R^m$ is the residual signal. L_i is the gain to be designed, and $sat_{\sigma(t)}(*)$ is a symmetric vector saturation function with variable non-negative saturation limits $\sigma(t) = (\sigma_1(t), \sigma_2(t), \ldots, \sigma_m(t)) \in R^m_{\geq 0}$ defined for each $a = (a_1, a_2, \ldots, a_m) \in R^m$ as:

$$sat_{\sigma(t)}(a) = \begin{bmatrix} sat_{\sigma_1(t)}(a_1) \\ sat_{\sigma_2(t)}(a_2) \\ \vdots \\ sat_{\sigma_m(t)}(a_m) \end{bmatrix} \tag{7.4}$$

in which $sat_{\sigma_k(t)}(a_k)$ $(k = 1, 2, \ldots, m)$ is the standard scalar symmetric saturation function, and the dynamics of the saturation limits $\sigma(t)$ is accomplished by the following equation:

$$\dot{\bar{\sigma}}(t) = -\lambda\bar{\sigma}(t) + (y(t) - \hat{y}(t))^T R(y(t) - \hat{y}(t)), \bar{\sigma}(t) \in R^m_{\geq 0} \tag{7.5}$$

$$\sigma_k(t) = \sqrt{\bar{\sigma}(t)/w_k}, k = 1, 2, \ldots, m \tag{7.6}$$

where $\lambda > 0$, $R > 0$ and $w_k > 0$ $(k = 1, 2, \ldots, m)$ are the parameters with appropriate dimensions. Moreover, $\bar{\sigma}(t) \in R^m_{\geq 0}$ implies that the Eq. (7.6) is always established.

The new fuzzy fault detection observer (7.3) is designed to detect the faults, and the adaptive saturation of output errors enable it to copy with the outliers, thereby preventing possible outliers from directly reaching the residual outputs and the error dynamics. Therefore, the advantage of this type of fault detection observer is to reduce the effect of the possible outliers in the measurement on the estimation errors and the residual signals. Thus, false alarms caused by the outliers can be effectively avoided. Moreover, the saturation level must be dynamically adjusted to ensure the exponentially stability of the error systems.

Refer to (7.2) and (7.3), define $\tilde{x}(t) = x(t) - \hat{x}(t)$, $e = \left[x^T(t)\ \tilde{x}^T(t)\right]^T$, and let $\bar{r}(t) = r(t) - Mf(t)$ be the fault estimation error, in which $M = \left[I \cdots I\right]^T_{m\times u}$. Then, the fault detection system are constructed, as shown below:

$$\begin{aligned}
\dot{e}(t) &= \bar{A}_{h\hat{h}}e(t) + \bar{D}_{h\hat{h}}v(t) + \bar{F}_{h\hat{h}}f(t) + \bar{E}_{h\hat{h}}q(t)\\
\bar{r}(t) &= \bar{C}e(t) - q(t) - Mf(t)\\
\dot{\bar{\sigma}}(t) &= -\lambda\bar{\sigma}(t) + (y(t) - \hat{y}(t))^T R(y(t) - \hat{y}(t)), \bar{\sigma}(t) \in R^m_{\geq 0}\\
q(t) &= dz_{\sigma(t)}(\bar{C}e(t))\\
&= \bar{C}e(t) - sat_{\sigma(t)}(\bar{C}e(t))
\end{aligned} \tag{7.7}$$

in which

$$\bar{A}_{h\hat{h}} = \begin{bmatrix} A_h & 0 \\ A_h - A_{\hat{h}} & A_{\hat{h}} - L_{\hat{h}}C \end{bmatrix}, \bar{D}_{h\hat{h}} = \begin{bmatrix} D_h \\ D_h \end{bmatrix}, \bar{F}_{h\hat{h}} = \begin{bmatrix} F_h \\ F_h \end{bmatrix}, \bar{E}_{h\hat{h}} = \begin{bmatrix} 0 \\ L_{\hat{h}} \end{bmatrix},$$
$$\bar{C} = \begin{bmatrix} 0 & C \end{bmatrix}$$

The residual signal $\bar{r}(t)$ is evaluated by the residual evaluation function established as follows:

$$J(t) = \left\{ \frac{1}{t} \int_0^t \bar{r}^T(\tau)\bar{r}(\tau)d\tau \right\}^{\frac{1}{2}} \tag{7.8}$$

The threshold is defined as:

$$J_{th} = \sup_{f(t)=0} J(t) \tag{7.9}$$

The generation of alarm can be described as follows:

$$\begin{cases} J(k) > J_{th} \Rightarrow \textit{with fault} \Rightarrow \textit{alarm} \\ J(k) \leq J_{th} \Rightarrow \textit{no fault} \end{cases}$$

Now, based on the above analysis, the purpose of this chapter is summarized as the following problem:

Problem: In order to study the problem of fuzzy fault detection observer design for T-S fuzzy systems with measurement outliers, the new fuzzy fault detection observer (7.3) is proposed to detect the faults. Thus, the fuzzy fault detection observer design problem will be converted to the following multi-objective requirements.

(S.1) *Exponentially Stability:* the fault detection system (7.7) with $v(t) = 0$ and $f(t) = 0$ is exponentially stable.

(S.2) H_∞ Performance: the fault detection system (7.7) satisfies the following inequality:

$$\|e(t)\|_2^2 \leq \delta_1^2 \|v(t)\|_2^2 \tag{7.10}$$

$$\|e(t)\|_2^2 \leq \delta_2^2 \|f(t)\|_2^2 \tag{7.11}$$

for all $v(t) \in L_2[0, \infty)$ and $f(t)$, in which δ_1 and δ_2 are the H_∞ performance bounds.

7.3 Fuzzy Fault Detection Observer With Adaptive Saturation of Output Errors Design

7.3.1 Stability Analysis

In this section, exponentially stability conditions of the fault detection system (7.7) with $v(t) = 0$ and $f(t) = 0$ will be constructed. In order to obtain the main results, the fault detection system (7.7) is rewritten as the following form:

$$\begin{aligned}
\dot{e}(t) &= \bar{A}_{h\hat{h}} e(t) + \bar{E}_{h\hat{h}} q(t) \\
\dot{\bar{\sigma}}(t) &= -\lambda \bar{\sigma}(t) + (y(t) - \hat{y}(t))^T R (y(t) - \hat{y}(t)), \bar{\sigma}(t) \in R^m_{\geq 0} \\
q(t) &= dz_{\sigma(t)}(\bar{C} e(t)) \\
&= \bar{C} e(t) - sat_{\sigma(t)}(\bar{C} e(t))
\end{aligned} \tag{7.12}$$

Next, the main results will be given. The gains of the fuzzy fault detection observer are presented according to some convex constraints, where the first constraint (7.13) guarantees that the interaction between adaptive parameters λ and R can make the growth of $\bar{\sigma}(t)$ fast enough, so as to prevent the divergence of the appropriate Lyapunov function $e^T(t) P_{\hat{h}} e(t)$.

$$He \begin{bmatrix} (\bar{1,1}) & P_{\hat{h}} E_{h\hat{h}} \\ U\bar{C} & -U \end{bmatrix} < 0 \tag{7.13}$$

where $He(\star) = \star + \star^T$, $(\bar{1,1}) = P_{\hat{h}} \bar{A}_{h\hat{h}} + \frac{1}{2}(\lambda P_{\hat{h}} - \bar{C}^T R \bar{C})$, $P_{\hat{h}} > 0$, and $U > 0$ is a diagonal matrix, which is used to increase the additional degrees of freedom.

The second constraint (7.14) is used to guarantee that, once $\bar{\sigma}(t)$ is sufficient to "capture" the errors, then $e(t)$ and $\sigma(t)$ will converge to zero.

$$He\begin{bmatrix} (1,1) & P_{\hat{h}}\bar{E}_{h\hat{h}} \\ W\bar{C}+Y & -W \end{bmatrix} < 0 \tag{7.14}$$

and

$$\begin{bmatrix} P_{\hat{h}} & Y_k^T \\ Y_k & w_k \end{bmatrix} \geq 0, \quad k \in (1,2,\ldots,m) \tag{7.15}$$

where $(1,1) = P_{\hat{h}}\bar{A}_{h\hat{h}}$, $Y = WH$, w_k represents the kth diagonal element of W, and Y_k is defined as the kth row of matrix Y.

Then, the main results are summarized as Theorem 7.1, which is shown as follows.

Theorem 7.1 *Consider the T-S fuzzy system (7.1). Given constant $\lambda > 0$, the fault detection system (7.7) is exponentially stable, if there exist matrices $P_j > 0$, L_j, $j = 1,2,\ldots,r$, $U > 0$, $R > 0$ and $W > 0$ such that $\dot{P}_{\hat{h}} \leq 0$ and the inequalities (7.13)–(7.15) are satisfied.*

Proof Theorem 7.1 is proved by introducing a novel non-differentiable Lyapunov function, and before the Lyapunov function is given, it is known that both (7.13) and (7.14) are strict, and therefore, a small enough scalar $\varepsilon > 0$ can be found to satisfy the following inequalities:

$$He\begin{bmatrix} (\bar{1,1}) & P_{\hat{h}}\bar{E}_{h\hat{h}} \\ U\bar{C} & -U \end{bmatrix} < -2\varepsilon \tag{7.16}$$

$$He\begin{bmatrix} P_{\hat{h}}\bar{A}_{h\hat{h}} & P_{\hat{h}}\bar{E}_{h\hat{h}} \\ W\bar{C}+WH & -W \end{bmatrix} < -2\varepsilon \tag{7.17}$$

Then, the Lyapunov functional is established, as shown below:

$$V(e(t),\bar{\sigma}(t)) = \underbrace{e^T(t)P_{\hat{h}}e(t) + \xi\bar{\sigma}(t)}_{V_1(e(t),\bar{\sigma}(t))} + \underbrace{\eta\max\{e^T(t)P_{\hat{h}}e(t) - \bar{\sigma}(t), 0\}}_{V_2(e(t),\bar{\sigma}(t))} \tag{7.18}$$

where $P_{\hat{h}} = \sum_{j=1}^{r} h_j(z(t))P_j$, ξ and η are positive scalars and are defined in the later.

Note that the Lyapunov functional $V(e(t),\bar{\sigma}(t))$ is non-differentiable in the zero measure set, where $e^T(t)P_{\hat{h}}e(t) - \bar{\sigma}(t) = 0$. Next, Theorem 7.1 is proved by constructing the decrease of $V(e(t),\bar{\sigma}(t))$ wherever it is differentiable, and two cases will be discussed, as shown below.

Case 1: $e^T(t)P_{\hat{h}}e(t) < \bar{\sigma}(t)$

First define $H = W^{-1}Y$, and use the Schur complement to (7.15), by performing the congruence transformation $e(t)$ to obtain $e^T(t)P_{\hat{h}}e(t) - w_k e^T(t)H_k^T H_k e(t) \geq 0$. Then, together with the condition $e^T(t)P_{\hat{h}}e(t) < \bar{\sigma}(t)$, one has:

$$w_k|H_k e(t)|^2 \le e^T(t)P_{\hat{h}}e(t) \le \bar{\sigma}(t) = w_k\sigma_k^2(t), \;\; k \in \{1, 2, \ldots, m\} \tag{7.19}$$

Moreover, it is easy to know $dz_{\sigma_k}(H_k e(t)) = 0$ for all $k \in \{1, 2, \ldots, m\}$ from (7.19), that is to say, $dz_\sigma(He(t)) = 0$. Based on this property, the following generalized sector conditions introduced by Refs. [17, 18] and rewritten by Ref. [19] can be constructed:

$$q^T(t)W\left(\bar{C}e(t) + He(t) - q(t)\right) \ge 0 \tag{7.20}$$

By performing the congruence transformation $(e(t), q(t))$ to (7.17), one has:

$$\begin{aligned} &2e^T(t)P_{\hat{h}}\left(\bar{A}_{h\hat{h}}e(t) + \bar{E}_{h\hat{h}}q(t)\right) + 2q^T(t)W\left(\bar{C}e(t) + He(t) - q(t)\right) \\ \le& -2\varepsilon|e(t)|^2 \end{aligned} \tag{7.21}$$

Define $\xi = \frac{\varepsilon}{|\bar{C}^T R\bar{C}|}$, and then calculate the derivative of $V(e(t), \bar{\sigma}(t))$, one has

$$\begin{aligned} \dot{V}(e(t), \bar{\sigma}(t)) &= 2e^T(t)P_{\hat{h}}\left(\bar{A}_{h\hat{h}}e(t) + \bar{E}_{h\hat{h}}q(t)\right) + e^T(t)\dot{P}_{\hat{h}}e(t) + \xi\dot{\bar{\sigma}}(t) \\ &= 2e^T(t)P_{\hat{h}}\left(\bar{A}_{h\hat{h}}e(t) + \bar{E}_{h\hat{h}}q(t)\right) + e^T(t)\dot{P}_{\hat{h}}e(t) \\ &\quad + \xi\left(-\lambda\bar{\sigma}(t) + e^T(t)\bar{C}^T R\bar{C}e(t)\right) \end{aligned} \tag{7.22}$$

Then, assume $\dot{P}_{\hat{h}} \le 0$, one has:

$$\begin{aligned} \dot{V}(e(t), \bar{\sigma}(t)) &= 2e^T(t)P_{\hat{h}}\left(\bar{A}_{h\hat{h}}e(t) + \bar{E}_{h\hat{h}}q(t)\right) + e^T(t)\dot{P}_{\hat{h}}e(t) + \xi\dot{\bar{\sigma}}(t) \\ &\le -2q^T(t)W(\bar{C}e(t) + He(t) - q(t)) - 2\varepsilon|e(t)|^2 \\ &\quad + \xi(-\lambda\bar{\sigma}(t) + e^T(t)\bar{C}^T R\bar{C}e(t)) \\ &\le -2\varepsilon|e(t)|^2 + \varepsilon|e(t)|^2 - \lambda\xi\bar{\sigma}(t) \\ &\le -\varepsilon|e(t)|^2 - \lambda\xi\bar{\sigma}(t) \end{aligned} \tag{7.23}$$

Case 2: $e^T(t)P_{\hat{h}}e(t) > \bar{\sigma}(t)$.

In this case, the maximum term in expression $V_2(e(t), \bar{\sigma}(t))$ is always the first one. By performing the congruence transformation $(e(t), q(t))$ to (7.16), one has:

$$\begin{aligned} &2e^T(t)P_{\hat{h}}\left(\bar{A}_{h\hat{h}}e(t) + \bar{E}_{h\hat{h}}q(t)\right) + 2q^T(t)U\left(\bar{C}e(t) - q(t)\right) \\ &+ \lambda e^T(t)P_{\hat{h}}e(t) - e^T(t)\bar{C}^T R\bar{C}e(t) \\ \le& -2\varepsilon|e(t)|^2 - 2\varepsilon|q(t)|^2 \end{aligned} \tag{7.24}$$

and then

$$\begin{aligned} \dot{V}_2(e(t), \bar{\sigma}(t)) &= 2e^T(t)P_{\hat{h}}\dot{e}(t) + e^T(t)\dot{P}_{\hat{h}}e(t) - \dot{\bar{\sigma}}(t) \\ &= 2e^T(t)P_{\hat{h}}\left(\bar{A}_{h\hat{h}}e(t) + \bar{E}_{h\hat{h}}q(t)\right) + e^T(t)\dot{P}_{\hat{h}}e(t) - \dot{\bar{\sigma}}(t) \end{aligned}$$

$$\begin{aligned}
&\leq 2e^T(t)P_{\hat{h}}\left(\bar{A}_{h\hat{h}}e(t)+\bar{E}_{h\hat{h}}q(t)\right)-e^T(t)\bar{C}^TR\bar{C}e(t)+\lambda\bar{\sigma}(t)\\
&\leq -2q^T(t)U(\bar{C}e(t)-q(t))-2\varepsilon(|e(t)|^2+|q(t)|^2)\\
&\leq -2\varepsilon(|e(t)|^2+|q(t)|^2)
\end{aligned} \tag{7.25}$$

where $\bar{\sigma}(t) < e^T(t)P_{\hat{h}}e(t)$ is used in the third step, and $q^T(t)U(\bar{C}e(t)+He(t)-q(t)) \geq 0$ with $H=0$ is used in the fourth step.

In addition, define $\eta = \frac{|P_1|}{2\varepsilon}$, and then one has:

$$\begin{aligned}
\dot{V}(e(t),\bar{\sigma}(t)) &= 2e^T(t)P_{\hat{h}}\dot{e}(t)+\xi\dot{\bar{\sigma}}(t)+\eta\dot{V}_2(e(t),\bar{\sigma}(t))\\
&\leq -2\varepsilon|e(t)|^2-\lambda\xi\bar{\sigma}(t)-2\eta\varepsilon(|e(t)|^2+|q(t)|^2)\\
&\leq -\varepsilon|e(t)|^2-\lambda\xi\bar{\sigma}(t)+2|P_1||e(t)||q(t)|-|P_1|(|e(t)|^2+|q(t)|^2)\\
&\leq -\varepsilon|e(t)|^2-\lambda\xi\bar{\sigma}(t)\\
&\leq -\varepsilon|e(t)|^2-\lambda\xi\sum_{k=1}^{m}w_k|\sigma(t)|^2
\end{aligned} \tag{7.26}$$

Based on the above analysis in Cases 1 and 2, it can be concluded that the fault detection system (7.7) is exponentially stable. The proof is completed. □

Remark 7.1 The inequality (7.13) ensures that the interaction between the adaptive parameters λ and R can make $\bar{\sigma}(t)$ grow fast enough, thus avoiding the appropriate Lyapunov function $V_1(e(t),\bar{\sigma}(t))$ divergence.

Remark 7.2 Similar to the Ref. [16], in order to obtain the main results of this chapter, a Lyapunov functional that is non-differentiable is established. By constructing the decrease of $V(e(t),\bar{\sigma}(t))$ wherever it is differentiable, a sufficient condition for the fault detection system (7.12) to satisfy the performance $(S.1)$ is summarized as Theorem 7.1.

Remark 7.3 The problems of sensor saturation are considered in Refs. [20, 21], in which the saturation level is fixed. Different from the existing results, dynamic saturation level is adopted in this chapter, and both the variable $e(t)$ and the variable $\bar{\sigma}(t)$ maintain the global exponential stability. Moreover, a sufficient condition is established to make the index of error $e(t)$ and $\sigma(t)$ exponentially converge to zero.

7.3.2 H_∞ Disturbance Attenuation

In this section, H_∞ performance conditions of the fault detection system (7.7) are given. Before giving the main results, the considered fault detection system is shown below:

$$\dot{e}(t) = \bar{A}_{h\hat{h}}e(t)+\bar{D}_{h\hat{h}}v(t)+\bar{E}_{h\hat{h}}q(t)$$

$$\begin{aligned} r(t) &= \bar{C}e(t) - q(t) \\ \dot{\bar{\sigma}}(t) &= -\lambda\bar{\sigma}(t) + (y(t) - \hat{y}(t))^T R(y(t) - \hat{y}(t)), \bar{\sigma}(t) \in R^m_{\geq 0} \\ q(t) &= dz_{\sigma(t)}(\bar{C}e(t)) \\ &= \bar{C}e(t) - sat_{\sigma(t)}(\bar{C}e(t)) \end{aligned} \tag{7.27}$$

Next, according to the Theorem 7.1, a sufficient condition for fault detection system (7.27) to satisfy the H_∞ performance condition can be easily summarized as Theorem 7.2, which is shown as follows.

Theorem 7.2 *Consider the T-S fuzzy system (7.1). Given constant $\lambda > 0$, the fault detection system (7.27) satisfies the exponential stability $(S.1)$ and the H_∞ performance (7.10), if there exist matrices $P_j > 0$, L_j, $j = 1, 2, \ldots, r$, $U > 0$, $R > 0$, $W > 0$ and $\delta_1 > 0$ such that $\dot{P}_{\hat{h}} \leq 0$ and the following inequalities are satisfied:*

$$He\begin{bmatrix} (1,1)_1 & P_{\hat{h}}\bar{E}_{h\hat{h}} - \bar{C}^T & P_{\hat{h}}\bar{D}_{h\hat{h}} \\ W\bar{C} + WH & -W + \frac{1}{2}I & 0 \\ 0 & 0 & -\frac{1}{2}\delta_1^2 I \end{bmatrix} < 0 \tag{7.28}$$

$$He\begin{bmatrix} (1,1)_2 & P_{\hat{h}}\bar{E}_{h\hat{h}} - \bar{C}^T & P_{\hat{h}}\bar{D}_{h\hat{h}} \\ U\bar{C} & -U + \frac{1}{2}I & 0 \\ 0 & 0 & -\frac{1}{2}\delta_1^2 I \end{bmatrix} < 0 \tag{7.29}$$

$$\begin{bmatrix} P_{\hat{h}} & Y_k^T \\ Y_k & w_k \end{bmatrix} \geq 0 \tag{7.30}$$

where $(1,1)_1 = P_{\hat{h}}\bar{A}_{h\hat{h}} + \frac{1}{2}\bar{C}^T\bar{C}$, $(1,1)_2 = P_{\hat{h}}\bar{A}_{h\hat{h}} + \frac{1}{2}(\lambda P_{\hat{h}} - \bar{C}^T R\bar{C}) + \frac{1}{2}\bar{C}^T\bar{C}$, *$Y = WH$.*

Proof The non-differentiable Lyapunov functional is established for the fault detection system (7.27), as shown below:

$$V(e(t), \bar{\sigma}(t)) = \underbrace{e^T(t)P_{\hat{h}}e(t) + \xi\bar{\sigma}(t)}_{V_1(e(t),\bar{\sigma}(t))} + \underbrace{\eta\max\{e^T(t)P_{\hat{h}}e(t) - \bar{\sigma}(t), 0\}}_{V_2(e(t),\bar{\sigma}(t))} \tag{7.31}$$

where $P_{\hat{h}} > 0$ is the same as that of (7.18).

The time-derivative of $V(e(t), \bar{\sigma}(t))$ is calculated along the trajectory of the fault detection system (7.27), as follows:

$$\dot{V}(e(t), \sigma(t)) = \left\{ \begin{cases} 2e^T(t)P_{\hat{h}}\dot{e}(t) + e^T(t)\dot{P}_{\hat{h}}e(t) + \xi\dot{\bar{\sigma}}(t), \\ e^T(t)P_{\hat{h}}e(t) < \bar{\sigma}(t) \\ 2e^T(t)P_{\hat{h}}\dot{e}(t) + e^T(t)\dot{P}_{\hat{h}}e(t) + \xi\dot{\bar{\sigma}}(t) + \eta\dot{V}_2(e(t), q(t)), \\ e^T(t)P_{\hat{h}}e(t) > \bar{\sigma}(t) \end{cases} \right. \tag{7.32}$$

where $\dot{V}_2(e(t), \sigma(t)) = 2e^T(t)P_{\hat{h}}\dot{e}(t) + e^T(t)\dot{P}_{\hat{h}}e(t) - \dot{\bar{\sigma}}(t)$.

In addition, define the following equation:

$$J(t) = \dot{V}(e(t), \sigma(t)) + r^T(t)r(t) - \delta_1^2 v^T(t)v(t) \tag{7.33}$$

Similar to the proof of Theorem 7.1, two cases need to be considered, one of which is:

Case 1: $e^T(t)P_{\hat{h}}e(t) < \bar{\sigma}(t)$.

Also define $\xi = \frac{\varepsilon}{|C^T RC|}$, and then

$$\begin{aligned}
J(t) &= \dot{V}(e(t), \sigma(t)) + r^T(t)r(t) - \delta_1^2 v^T(t)v(t) \\
&\leq 2e^T(t)P_{\hat{h}}\left(\bar{A}_{h\hat{h}}e(t) + \bar{E}_{h\hat{h}}q(t) + \bar{D}_{h\hat{h}}v(t)\right) + e^T(t)\dot{P}_{\hat{h}}e(t) \\
&\quad + \xi\dot{\bar{\sigma}}(t) + r^T(t)r(t) - \delta_1^2 v^T(t)v(t) \\
&\leq -2q^T(t)W(\bar{C}e(t) + He(t) - q(t)) - 2\varepsilon|e(t)|^2 \\
&\quad + \xi(-\lambda\bar{\sigma}(t) + e^T(t)\bar{C}^T R\bar{C}e(t)) \\
&\leq (-2\varepsilon + \xi|\bar{C}^T R\bar{C}|)|e(t)|^2 - \lambda\xi\bar{\sigma}(t) \\
&\leq -\varepsilon|e(t)|^2 - \lambda\xi\bar{\sigma}(t)
\end{aligned} \tag{7.34}$$

Case 2: $e^T(t)P_{\hat{h}}e(t) > \bar{\sigma}(t)$.

First, the time-derivative of $V_2(e(t), \bar{\sigma}(t))$ is calculated as follows:

$$\begin{aligned}
\dot{V}_2(e(t), \bar{\sigma}(t)) &= 2e^T(t)P_{\hat{h}}\dot{e}(t) + e^T(t)\dot{P}_{\hat{h}}e(t) - \dot{\bar{\sigma}}(t) \\
&= 2e^T(t)P_{\hat{h}}\left(\bar{A}_{h\hat{h}}e(t) + \bar{E}_{h\hat{h}}q(t) + \bar{D}_{h\hat{h}}v(t)\right) + e^T(t)\dot{P}_{\hat{h}}e(t) - \dot{\bar{\sigma}}(t) \\
&\leq 2e^T(t)P_{\hat{h}}\left(\bar{A}_{h\hat{h}}e(t) + \bar{E}_{h\hat{h}}q(t) + \bar{E}_{h\hat{h}}v(t)\right) \\
&\quad - e^T(t)\bar{C}^T R\bar{C}e(t) + \lambda\bar{\sigma}(t) \\
&\leq -2q^T(t)U(\bar{C}e(t) - q(t)) - 2\varepsilon(|e(t)|^2 + |q(t)|^2) \\
&\leq -2\varepsilon(|e(t)|^2 + |q(t)|^2)
\end{aligned} \tag{7.35}$$

Then, also define $\eta = \frac{|P_1|}{2\varepsilon}$, one has:

$$\begin{aligned}
J(t) &= \dot{V}(e(t), \bar{\sigma}(t)) + r^T(t)r(t) - \delta^2 v^T(t)v(t) \\
&\leq 2e^T(t)P_{\hat{h}}\dot{e}(t) + e^T(t)\dot{P}_{\hat{h}}e(t) + \xi\dot{\bar{\sigma}}(t) + \eta\dot{V}_2(e(t), \bar{\sigma}(t)) \\
&\quad + r^T(t)r(t) - \delta_1^2 v^T(t)v(t) \\
&\leq (-2\varepsilon + \xi|\bar{C}^T R\bar{C}|)|e(t)|^2 - \lambda\xi\bar{\sigma}(t) - 2\eta(\varepsilon)(|e(t)|^2 + |q(t)|^2) \\
&\leq -\varepsilon|e(t)|^2 - \lambda\xi\bar{\sigma}(t) + 2|P_1||e(t)||q(t)| - |P_1|(|e(t)|^2 + |q(t)|^2) \\
&\leq -\varepsilon|e(t)|^2 - \lambda\xi\bar{\sigma}(t) \\
&\leq -\varepsilon|e(t)|^2 - \lambda\xi\sum_{k=1}^{m} w_k|\sigma(t)|^2
\end{aligned} \tag{7.36}$$

Thus, it can be seen that the H_∞ performance condition (7.10) of the fault detection system is satisfied. The proof is completed. □

The proposed method can be reduced to the existing ones in Refs. [9, 24] by removing the adaptive saturation of output errors, which is presented as the following corollary.

$$\begin{aligned}\dot{e}(t) &= \bar{A}_{h\hat{h}}e(t) + \bar{D}_{h\hat{h}}v(t)\\ r(t) &= \bar{C}e(t).\end{aligned} \tag{7.37}$$

Corollary 7.1 *Consider the T-S fuzzy system (7.1). The fault detection system (7.37) satisfies the H_∞ performance (7.10), if there exist matrices $P_j > 0$ and L_j, $j = 1, 2, \ldots, r$ such that $\dot{P}_{\hat{h}} < 0$ and the following inequalities are satisfied.*

$$He\begin{bmatrix} P_{\hat{h}}\bar{A}_{h\hat{h}} + \frac{1}{2}\bar{C}^T\bar{C} & P_{\hat{h}}\bar{D}_{h\hat{h}} \\ 0 & -\frac{1}{2}\delta_1^2 I \end{bmatrix} < 0. \tag{7.38}$$

7.3.3 H_∞ Fault Attenuation

In this section, H_∞ performance conditions of the fault detection system (7.7) are constructed by considering the influence of fault on fault estimation error $\bar{r}(t)$. Before presenting the main results, the system (7.7) in the absence of disturbance can be constructed as follows:

$$\begin{aligned}\dot{e}(t) &= \bar{A}_{h\hat{h}}e(t) + \bar{F}_{h\hat{h}}f(t) + \bar{E}_{h\hat{h}}q(t)\\ \bar{r}(t) &= \bar{C}e(t) - q(t) - Mf(t)\\ \dot{\bar{\sigma}}(t) &= -\lambda\bar{\sigma}(t) + (y(t) - \hat{y}(t))^T R(y(t) - \hat{y}(t)), \bar{\sigma}(t) \in R^m_{\geq 0}\\ q(t) &= dz_{\sigma(t)}(\bar{C}e(t))\\ &= \bar{C}e(t) - sat_{\sigma(t)}(\bar{C}e(t))\end{aligned} \tag{7.39}$$

Next, according to Theorem 7.3, a sufficient condition for the fault detection system (7.39) to satisfy the H_∞ performance condition (7.11) is established as follows.

Theorem 7.3 *Consider the T-S fuzzy system (7.1). Given some constants $\delta_1 > 0$, and $\lambda > 0$, the fault detection system (7.39) satisfies the exponential stability $(S.1)$ and the H_∞ performance (7.11), if there exist matrices $P_j > 0$, L_j, $j = 1, 2, \ldots, r$, $U > 0$, $R > 0$, $W > 0$ and $\delta_2 > 0$ such that $\dot{P}_{\hat{h}} \leq 0$ and the following inequalities are satisfied.*

$$He\begin{bmatrix} \overline{(1,1)}_1 & P_{\hat{h}}\bar{E}_{h\hat{h}} - \bar{C}^T & P_{\hat{h}}\bar{F}_{h\hat{h}} - \bar{C}^T M \\ W\bar{C} + WH & -W + \frac{1}{2}I & M \\ 0 & 0 & -\frac{1}{2}\delta_2^2 I + \frac{1}{2}M^T M \end{bmatrix} < 0 \tag{7.40}$$

$$He\begin{bmatrix} \overline{(1,1)}_2 & P_{\hat{h}}\bar{E}_{h\hat{h}} - \bar{C}^T & P_{\hat{h}}\bar{F}_{h\hat{h}} - \bar{C}^T M \\ U\bar{C} & -U + \frac{1}{2}I & M \\ 0 & 0 & -\frac{1}{2}\delta_2^2 I + \frac{1}{2}M^T M \end{bmatrix} < 0 \tag{7.41}$$

$$\begin{bmatrix} P_{\hat{h}} & Y_k^T \\ Y_k & w_k \end{bmatrix} \geq 0 \tag{7.42}$$

where $\overline{(1,1)}_1 = P_{\hat{h}}\bar{A}_{h\hat{h}} + \frac{1}{2}\bar{C}^T\bar{C}$, $\overline{(1,1)}_2 = P_{\hat{h}}\bar{A}_{h\hat{h}} + \frac{1}{2}(\lambda P_{\hat{h}} - \bar{C}^T R\bar{C}) + \frac{1}{2}\bar{C}^T\bar{C}$, $Y = WH$.

Proof The main idea for proving Theorem 7.3 is the same as that of Theorem 7.2, which is omitted here for brevity. □

Remark 7.4 Measurement outliers can completely destroy the output measurements used to build the residual signals [16], which will lead to false alarms. Therefore, a new fuzzy fault detection observer is constructed by using the adaptive saturation of output errors to reduce the influence of measurement outliers on the estimation errors and the residual signals.

7.3.4 Algorithm

For (7.28), (7.29), (7.40) and (7.41), extract the fuzzy basis functions and define $P_s = \begin{bmatrix} P_{s11} & -P_{s22} \\ -P_{s22} & P_{s22} \end{bmatrix}$, let $P_{s22}L_j$ be a new variable K_{sj}, then the following inequalities are satisfied with $i, j, s = 1, 2, \ldots, r$, $k = 1, 2, \ldots, m$:

$$He\begin{bmatrix} (1,1)_1 & P_s\bar{E}_{ij} - \bar{C}^T & P_s\bar{D}_{ij} \\ W\bar{C} + WH & -W + \frac{1}{2}I & 0 \\ 0 & 0 & -\frac{1}{2}\delta_1^2 I \end{bmatrix} < 0 \tag{7.43}$$

$$He\begin{bmatrix} (1,1)_2 & P_s\bar{E}_{ij} - \bar{C}^T & P_s\bar{D}_{ij} \\ U\bar{C} & -U + \frac{1}{2}I & 0 \\ 0 & 0 & -\frac{1}{2}\delta_1^2 I \end{bmatrix} < 0 \tag{7.44}$$

$$He\begin{bmatrix} \overline{(1,1)}_1 & P_s\bar{E}_{ij} - \bar{C}^T & P_s\bar{F}_{ij} - \bar{C}^T M \\ W\bar{C} + WH & -W + \frac{1}{2}I & M \\ 0 & 0 & -\frac{1}{2}\delta_2^2 I + \frac{1}{2}M^T M \end{bmatrix} < 0 \tag{7.45}$$

$$He\begin{bmatrix} \overline{(1,1)}_2 & P_s\bar{E}_{ij} - \bar{C}^T & P_s\bar{F}_{ij} - \bar{C}^T M \\ U\bar{C} & -U + \frac{1}{2}I & M \\ 0 & 0 & -\frac{1}{2}\delta_2^2 I + \frac{1}{2}M^T M \end{bmatrix} < 0 \tag{7.46}$$

$$\begin{bmatrix} P_s & Y_k^T \\ Y_k & w_k \end{bmatrix} \geq 0 \tag{7.47}$$

where

$$(1,1)_1 = P_s\bar{A}_{ij} + \frac{1}{2}\bar{C}^T\bar{C}, Y = WH, (1,1)_2 = P_s\bar{A}_{ij} + \frac{1}{2}(\lambda P_s - \bar{C}^T R\bar{C}) + \frac{1}{2}\bar{C}^T\bar{C},$$
$$(1,\bar{1})_1 = P_s\bar{A}_{ij} + \frac{1}{2}\bar{C}^T\bar{C}, (1,\bar{1})_2 = P_s\bar{A}_{ij} + \frac{1}{2}(\lambda P_s - \bar{C}^T R\bar{C}) + \frac{1}{2}\bar{C}^T\bar{C}.$$

Then, the gains of the fault detection observer can be presented by the following algorithm.

Algorithm 1: Given $\delta_1 > 0$, by solving (7.43)–(7.47), a feasible solution to the fault detection observer (7.4) is presented as follows:

$$\begin{gathered} \min \delta_2 \\ s.t.\ (7.43), (7.44), (7.45), (7.46), (7.47) \\ i, j, s = 1, 2, \ldots, r, k = 1, 2, \ldots, m \end{gathered} \tag{7.48}$$

Then, the gains of the fault detection observer can be obtained as $L_j = P_{s22}^{-1}K_{sj}$ $(j = 1, 2, \ldots, r)$.

Remark 7.5 Similar to the Refs. [22, 23], it is easy to deal with the fuzzy Lyapunov functions and is omitted here.

Remark 7.6 The number of linear matrix inequalities given in this chapter and the ones obtained by adopting the methods in Refs. [9, 24] are $4r^3 + m$ and $2r$, respectively, and compared with the linear matrix inequalities obtained in the latter, each linear matrix inequality given in the former increases only one row and one column. Therefore, the complexity of the algorithm constructed in this paper is not high.

7.4 Numeral Examples

Two simulation examples are given in this section to demonstrate the effectiveness of the proposed new fuzzy fault detection observer design method.

7.4.1 Example 1

The problem of balancing an inverted pendulum on a cart in Ref. [25] is considered, as shown below:

$$\begin{aligned} \dot{x}_1 &= x_2 \\ \dot{x}_2 &= \frac{g\ sin(x_1) - aml/x_2^2\ sin(2x_1)/2 - a\ cos(x_1)u}{(4l)/3 - aml\ cos(x_1)^2} + w \end{aligned}$$

where x_1 represents the angle of the pendulum from the vertical, x_2 is the angular velocity, w represents the external disturbance, m and M are, respectively, the mass of the pendulum and the mass of the cart, $2l$ is the length of the pendulum. The corresponding parameters are given as: $g = 9.8\,\mathrm{m/g^2}, m = 2.0\,\mathrm{kg}, M = 8.0\,\mathrm{kg}, 2l = 1.0\,\mathrm{m}, a = \frac{1}{m+M}$. Then, by using the fuzzy modeling methods, assuming that the fault exists in the system and considering the static output feedback controller $u(t) = 300y(t)$, the above nonlinear system model is modeled in the form of T-S fuzzy model (7.1), where the membership functions are as follows:

$$h_1(z(t)) = \left(1 - \frac{1}{1+e^{-7(x_1-\frac{\pi}{4})}}\right)\left(\frac{1}{1+e^{-7(x_1+\frac{\pi}{4})}}\right)$$
$$h_2(z(t)) = 1 - h_1(z(t))$$

and the system matrices are given as follows:

$$A_1 = \begin{bmatrix} 0 & 1 \\ -35.6559 & -52.9500 \end{bmatrix}, A_2 = \begin{bmatrix} 0 & 1 \\ -10.7395 & -23.3700 \end{bmatrix},$$
$$D_1 = \begin{bmatrix} 1 \\ 1 \end{bmatrix}, D_2 = \begin{bmatrix} 1 \\ 1 \end{bmatrix}, F_1 = \begin{bmatrix} 0 \\ 1 \end{bmatrix}, F_2 = \begin{bmatrix} 1 \\ 1.5 \end{bmatrix}, C = \begin{bmatrix} 1 & 1 \end{bmatrix}$$

By solving (7.43), the fuzzy fault detection observer gain matrices are obtained, as shown below:

$$L_1 = \begin{bmatrix} 0.0877 \\ -2.6321 \end{bmatrix}, L_2 = \begin{bmatrix} -0.0780 \\ 2.3589 \end{bmatrix}, R = 180.6259, W = 6.2577, \ \dot{h}_1 \le 0$$
$$L_1 = \begin{bmatrix} 0.2494 \\ -0.4396 \end{bmatrix}, L_2 = \begin{bmatrix} -0.0220 \\ 0.1413 \end{bmatrix}, R = 38.5059, W = 32.6233, \ \dot{h}_1 > 0$$

In this chapter, the new fault detection observer (7.3) is designed for the T-S fuzzy system (7.1) to detect the faults, the fault detection system (7.7) needs to satisfy three performance conditions: exponential stability (S.1), H_∞ performance conditions (7.10) and (7.11), and the residual signals can suppress the outliers in the measurements. Therefore, in this example, the effectiveness of the proposed fault detection observer design method is verified in the following three cases.

Case I: In this case, exponential stability and H_∞ disturbance attenuation performance of the fault detection system (7.7) is illustrated. For this purpose, it is assumed that $f(t) = 0$ in the T-S fuzzy system (7.1), and the disturbance signal satisfies $w(t) = 0.4sin(0.3t)$ $(0 < t < 8)$.

The estimation errors of the systems are plotted in Fig. 7.1, which shows that the disturbance signal can be effectively suppressed by the proposed method in this chapter, and the estimation errors of the systems can converge to zero exponentially in the absence of disturbance. The trajectory $\bar{\sigma}$ in Case I is plotted in Fig. 7.2. From Fig. 7.2, it can be seen that when the errors change quickly, $\bar{\sigma}(t)$ can grow fast enough,

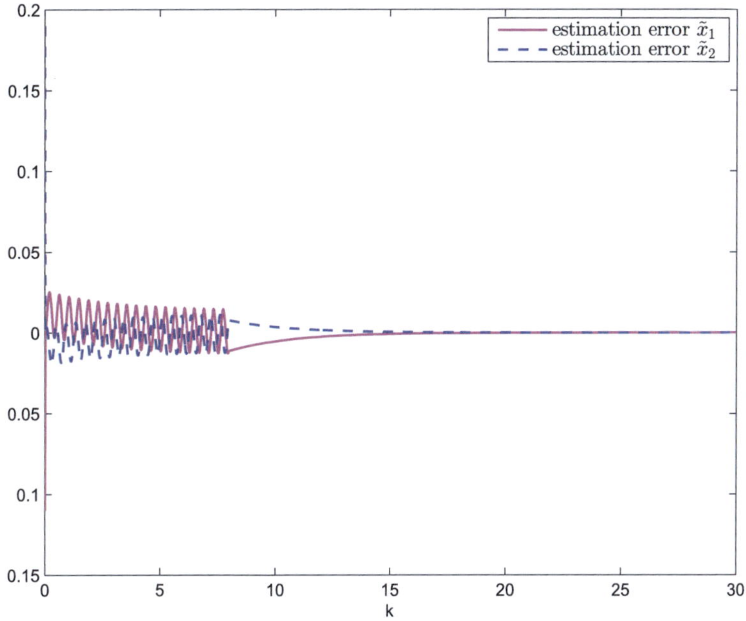

Fig. 7.1 Estimation error of the system in Case I

so as to prevent the divergence of the appropriate Lyapunov function $e^T(t)P_{\hat{h}}e(t)$, and when the errors converge to zero in case of $w(t) = 0$, $\sigma(t)$ will converge to zero.

Case II: In this case, the validity of the proposed fault detection approach in this chapter is verified. For this purpose, it is assumed that the fault is generated in the T-S fuzzy system and satisfies $f(t) = 1$ $(8 < t < 30)$, and the disturbance signal satisfies $w(t) = 0.4sin(0.3t)$ $(0 < t < 30)$.

The evaluation function $J_r(t)$ and fault detection threshold J_{th} are shown in Fig. 7.3, and it can be seen that $J_r(t)$ generated by the fuzzy fault detection observer satisfies $J_r(t) > J_{th}$ at $t = 11.84s$, thus displaying the fault is detected at $t = 11.84s$.

Case III: In this case, the validity of the constructed residual generator in this chapter is illustrated. For this purpose, it is assumed that the outliers in the measurements is a unit pulse signal and enters the system at $k = 8$, the fault signal satisfies $f(t) = 0$ $(0 < t < 30)$, and the disturbance signal satisfies $w(t) = 0.4sin(0.3t)$ $(0 < t < 30)$.

The evaluation function $J_r(t)$ and fault detection threshold J_{th} obtained by using the proposed approach are presented in Fig. 7.4, and the residual evaluation function $J_r(t)$ and fault detection threshold J_{th} generated in Refs. [8] are shown in Fig. 7.5. From Fig. 7.4, it can be seen that false alarms are not generated in the system in the case of outliers in the measurements, and a false alarm is generated in Fig. 7.5, which demonstrates the effectiveness of the proposed design method.

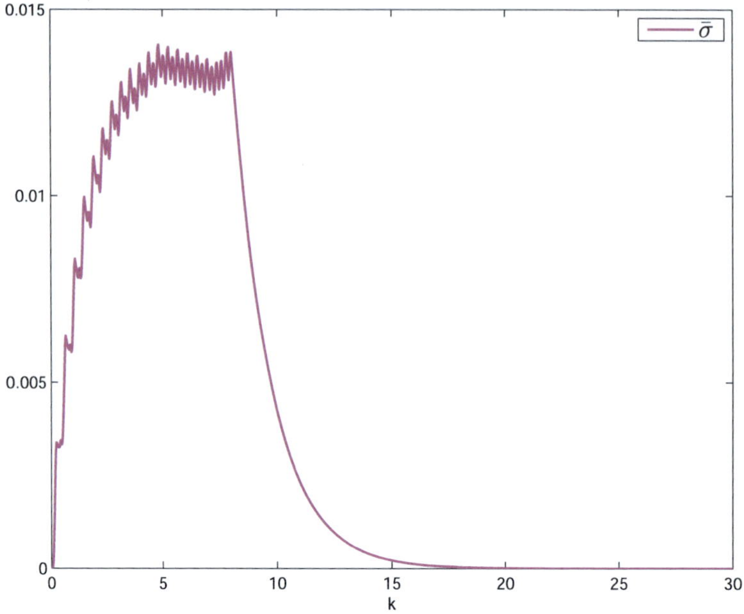

Fig. 7.2 History of the updating parameters $\bar{\sigma}$ in Case I

7.4.2 Example 2

In this example, the tunnel diode circuit in Ref. [26] is used to verify the effectiveness of the proposed method, in which the circuit diagram of tunnel diode circuit is shown in Fig. 2.1 of Chap. 2.

The parameter matrices comes from the Ref. [26] and are presented below:

$$A_1 = \begin{bmatrix} -0.1 & 50 \\ -1 & -10 \end{bmatrix}, A_2 = \begin{bmatrix} -4.6 & 50 \\ -1 & -10 \end{bmatrix}, D_1 = \begin{bmatrix} 0 \\ 1 \end{bmatrix},$$
$$D_2 = \begin{bmatrix} 0 \\ 1 \end{bmatrix}, F_1 = \begin{bmatrix} 0 \\ 1 \end{bmatrix}, F_2 = \begin{bmatrix} 1 \\ 0.5 \end{bmatrix}, C = \begin{bmatrix} 1 & 0 \end{bmatrix}$$

In addition, select the same membership functions as in the Ref. [26].

By solving (7.43), the fuzzy fault detection observer gains are obtained, as shown below:

$$L_1 = \begin{bmatrix} -11.8570 \\ 2.1835 \end{bmatrix}, L_2 = \begin{bmatrix} 3.8426 \\ -0.7185 \end{bmatrix}, R = 197.3352, W = 3.8554, \quad \dot{h}_1 \leq 0$$
$$L_1 = \begin{bmatrix} 4.8818 \\ 0.7270 \end{bmatrix}, L_2 = \begin{bmatrix} -1.8491 \\ -0.2543 \end{bmatrix}, R = 15.3154, W = 8.0771, \quad \dot{h}_1 > 0$$

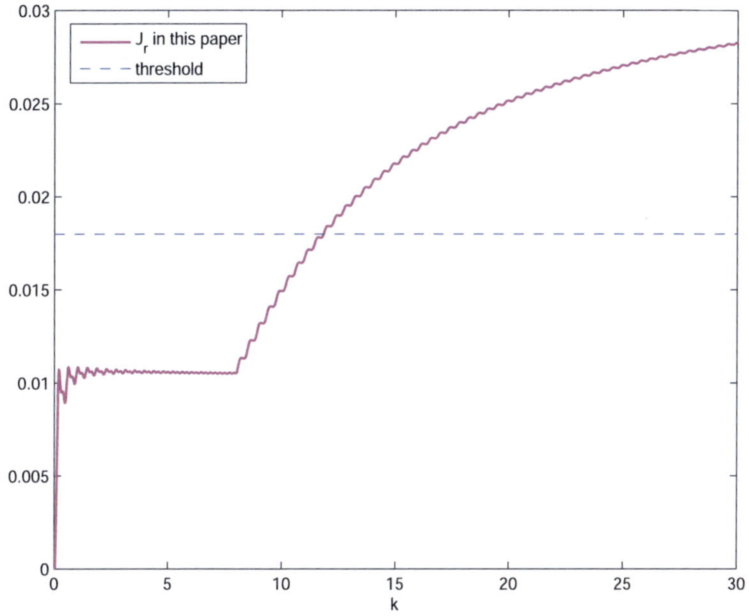

Fig. 7.3 Residual evaluation output $J(t)$ and threshold obtained by using the proposed fault detection method in Case II

Similar to Example 7.4.2, the effectiveness of the proposed method is illustrated in three cases, and the simulation results are presented in Figs. 7.6, 7.7, 7.8, 7.9 and 7.10.

Case I: Suppose $f(t) = 0$ in the T-S fuzzy system (7.1), and the disturbance signal satisfies $w(t) = 0.4sin(0.3t)$ $(0 < t < 4)$.

In Fig. 7.6, it is shown that the disturbance signal is effectively suppressed, and the estimation errors of the systems can converge to zero exponentially in the absence of disturbance. Moreover, it can be seen from Fig. 7.7 that $\bar{\sigma}$ can react quickly to the error signal $e(t)$.

Case II: Suppose $f(t) = 1$ $(4 < t < 30)$ in the T-S fuzzy system (7.1), and the disturbance signal satisfies $w(t) = 0.4sin(0.3t)$ $(0 < t < 30)$.

The evaluation function $J_r(t)$ and fault detection threshold J_{th} are shown in Fig. 7.8, and it is shown from Fig. 7.8 that the fault is detected at approximately $t = 4.65$.

Case III: Similar to Example 7.4.2, also assume the outliers in the measurements is a unit pulse signal and enters the system at $k = 4$, the fault signal satisfies $f(t) = 0$ $(0 < t < 30)$, and the disturbance signal satisfies $w(t) = 0.4sin(0.3t)$ $(0 < t < 30)$.

The residual evaluation function $J_r(t)$ and fault detection threshold J_{th} obtained by using the proposed method are presented in Fig. 7.9, and the evaluation function $J_r(t)$ and fault detection threshold J_{th} generated in Ref. [11] are shown in Fig. 7.10.

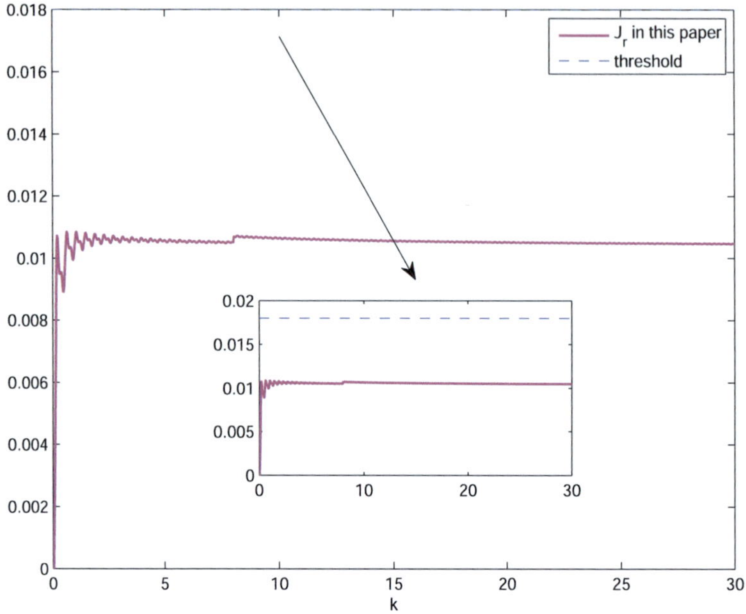

Fig. 7.4 Residual evaluation output $J(t)$ and threshold obtained by using the proposed fault detection method in Case III

From Figs. 7.9 and 7.10, it is shown that in the case of outliers in the measurements, false alarms using the method proposed in this chapter will not occur in the system, and false alarms will occur in the system using the method proposed in Ref. [11]. Therefore, the method designed in this chapter is effective.

7.5 Summary

In this chapter, the problem of fault detection observer design for T-S fuzzy systems with measurement outliers has been investigated. In order to avoid false alarms caused by measurement outliers, a new fuzzy fault detection observer is constructed by using the adaptive saturation of output errors. A suitable non-differentiable Lyapunov function is introduced to construct the exponential stability conditions and the H_∞ performance conditions of the fault detection systems. The design results of the fault detection observers are given based on the performance analysis conditions. Finally, the effectiveness of the proposed method is verified by the simulation examples.

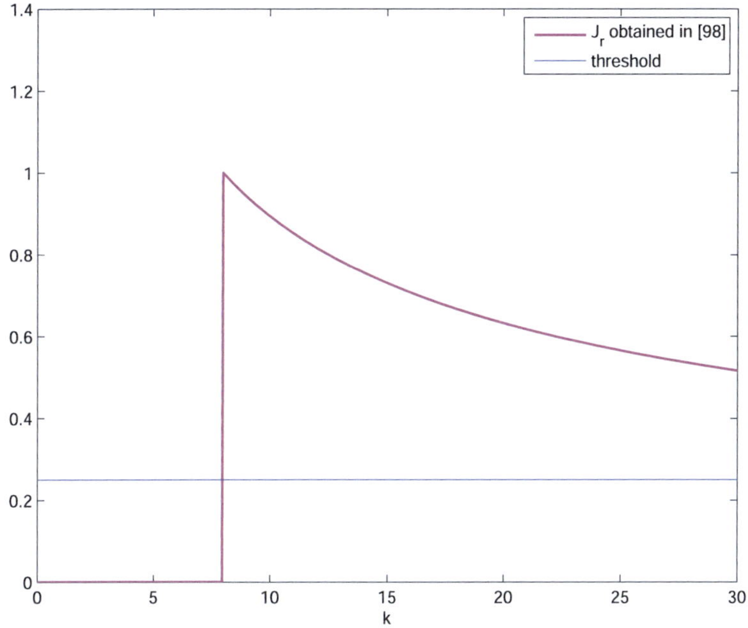

Fig. 7.5 Residual evaluation output $J(t)$ and threshold obtained in [8]

7.6 Notes

The outliers will completely destroy the output measurements used to build the residual signals, leading to the generation of false alarms, and there are few reports on handling measurement outliers in the research results of fault detection. Therefore, it is necessary to establish a new fuzzy fault detection observer design scheme for T-S fuzzy systems to reduce the influence of measurement outliers on the estimation errors and residual signals, which motivates this paper. In addition, it should be noted that the results presented in this chapter are mainly constructed based on the ones in Ref. [27].

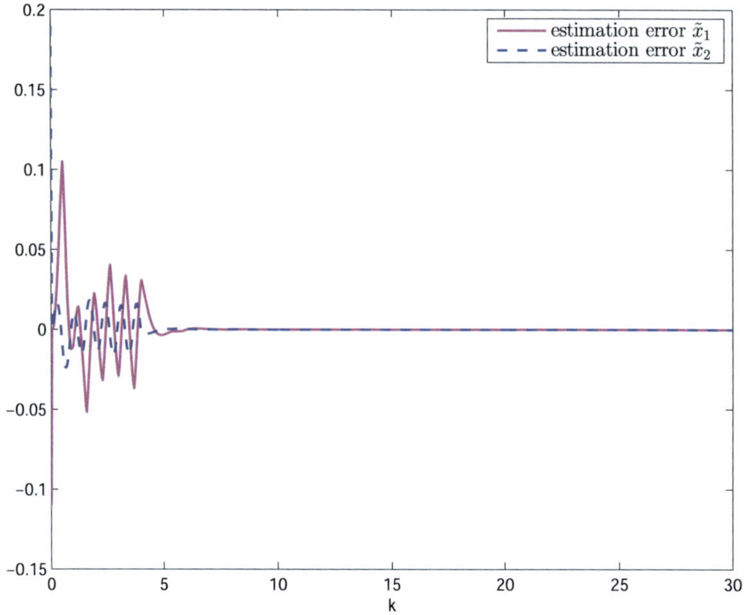

Fig. 7.6 Estimation error of the system in Case I

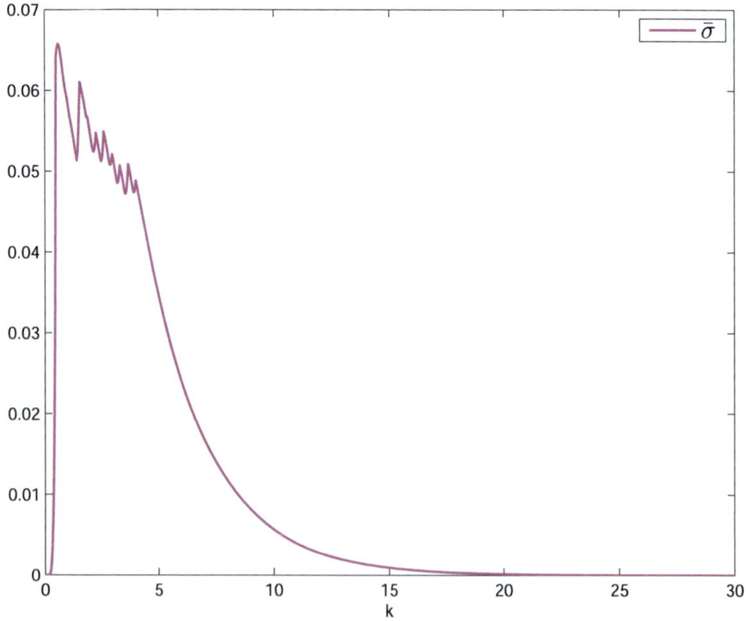

Fig. 7.7 History of the updating parameters $\bar{\sigma}$ in Case I

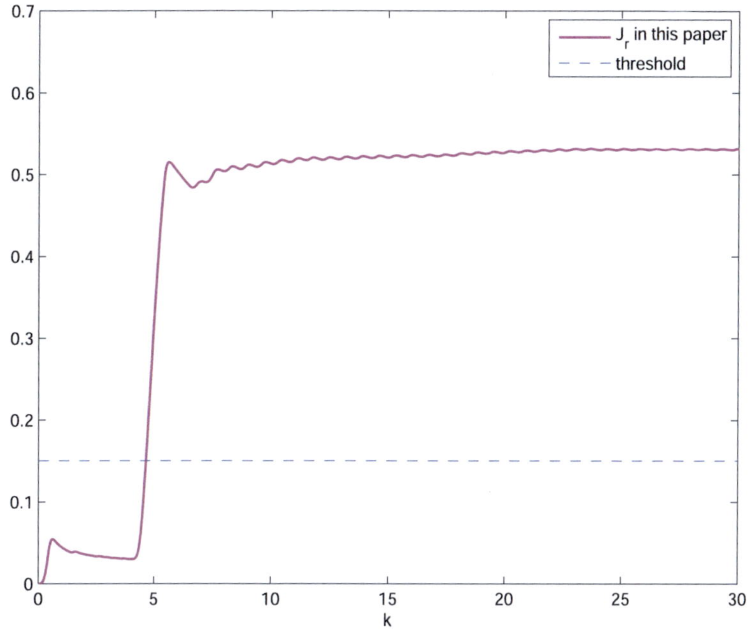

Fig. 7.8 Residual evaluation output $J(t)$ and threshold obtained by using the proposed fault detection method in Case II

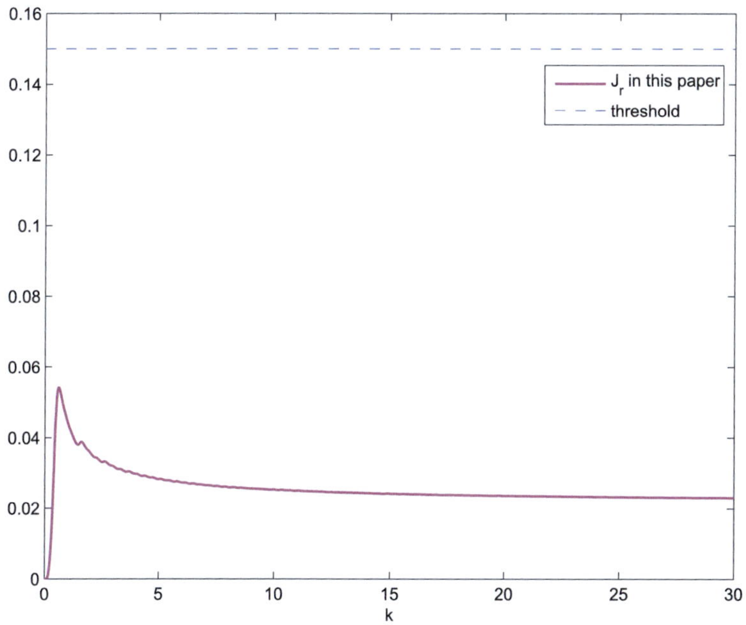

Fig. 7.9 Residual evaluation output $J(t)$ and threshold obtained by using the proposed fault detection method in Case III

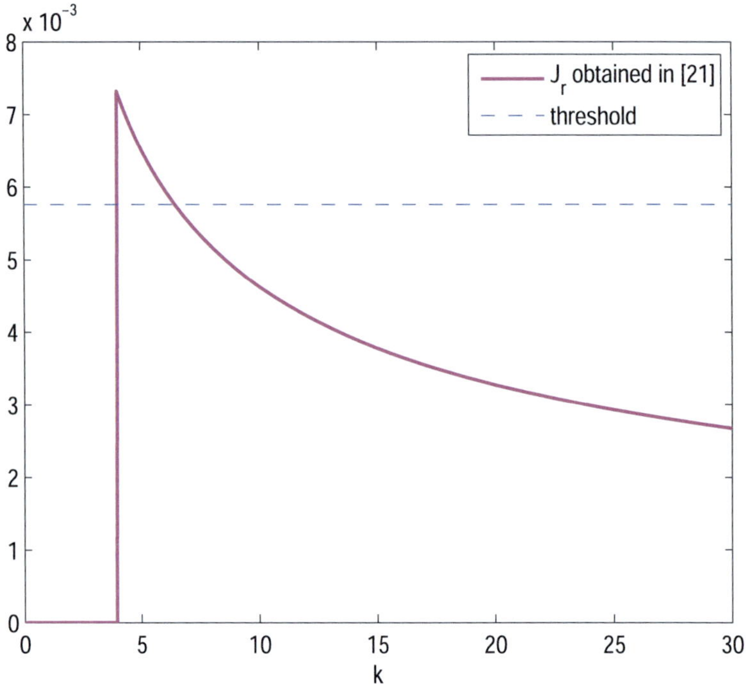

Fig. 7.10 Residual evaluation output $J(t)$ and threshold obtained in [11]

References

1. Chang XH, Li ZM, Park JH (2018) Fuzzy generalized H_2 filtering for nonlinear discrete-time systems with measurement quantization. IEEE Trans Syst Man Cybern: Syst 48(12):2419–2430
2. Dong JX, Wang YY, Yang GH (2009) Control synthesis of continuous-time T-S fuzzy systems with local nonlinear models. IEEE Trans Syst Man Cybern, Part B (Cybern) 39(5):1245–1258
3. Dong HL, Wang ZD, Ho D, Gao HJ (2010) Robust H_∞ fuzzy output-feedback control with multiple probabilistic delays and multiple missing measurements. IEEE Trans Fuzzy Syst 18(4):712–725
4. Wang XL, Yang GH (2018) Robust H_∞ filter design for discrete-time interconnected fuzzy systems with partially unknown membership functions and past output measurements. Neurocomputing 310:287–298
5. Wang XL, Yang GH (2019) Fault detection for T-S fuzzy systems with past output measurements. Fuzzy Sets Syst 365:98–115
6. Zhang ZH, Yang GH (2018) Interval observer-based fault detection in finite frequency domain for discrete-time fuzzy systems. Neurocomputing 310:38–45
7. Wu CW, Li HY, Lam HK, Karimi HR (2016) Fault detection for nonlinear networked systems based on quantization and dropout compensation: an interval type-2 fuzzy-model method. Neurocomputing 191:409–420
8. Li XJ, Yang GH (2014) Fault detection for T-S fuzzy systems with unknown membership functions. IEEE Trans Fuzzy Syst 22(1):139–152
9. Guo HF, Qiu JB, Tian H, Gao HJ (2014) Fault detection of discrete-time T-S fuzzy affine systems based on piecewise Lyapunov functions. J Franklin Inst 351(7):3633–3650

10. Chadli M, Abdo A, Ding SX (2013) H_{-}/H_{∞} fault detection filter design for discrete-time Takagi-Sugeno fuzzy system. Automatica 49(7):1996–2005
11. Pan YN, Yang GH (2018) Event-triggered fault detection filter design for nonlinear networked systems. IEEE Trans Syst Man Cybern 48(11):1851–1862
12. Gandhi MA, Mili L (2010) Robust Kalman filter based on a generalized maximum-likelihood-type estimator. IEEE Trans Signal Process 58(5):2509–2520
13. Gibbs RG (2013) New Kalman filter and smoother consistency tests. Automatica 49(10):3141–3144
14. Akkaya DA, Tiku ML (2008) Robust estimation in multiple linear regression model with non-Gaussian noise. Automatica 44(2):407–417
15. Alessandri A, Awawdeh M (2016) Moving-horizon estimation with guaranteed robustness for discrete-time linear systems and measurements subject to outliers. Automatica 67:85–93
16. Alessandri A, Zaccarian L (2018) Stubborn state observers for linear time-invariant systems. Automatica 88:1–9
17. Da SJMG, Tarbouriech S (2005) Anti-windup design with guaranteed regions of stability: an LMI-based approach. IEEE Trans Autom Control 50(1):106–111
18. Hu TS, Teel AR, Zaccarian L (2008) Anti-windup synthesis for linear control systems with input saturation: achieving regional nonlinear performance. Automatica 44(2):512–519
19. Tarbouriech S, Garcia G, Jr JMG da Silva, Queinnec I (2011) Stability and stabilization of linear systems with saturating actuators. Springer Science and Business Media
20. Yang FW, Li YM (2009) Set-membership filtering for systems with sensor saturation. Automatica 45(8):1896–1902
21. Wang ZD, Shen B, Liu XH (2012) H_{∞} filtering with randomly occurring sensor saturations and missing measurements. Automatica 48(3):556–562
22. Wang LK, Lam HK (2019) New stability criterion for continuous-time Takagi-Sugeno fuzzy systems with time-varying delay. IEEE Trans Cybern 49:1551–1556
23. Wang X, Park JH, She K, Zhong SM, Shi L (2019) Stabilization of chaotic systems with T-S fuzzy model and nonuniform sampling: a switched fuzzy control approach. IEEE Trans Fuzzy Syst 27(6):1263–1271
24. Li HY, Wu C, Yin S, Lam HK (2016) Observer-based fuzzy control for nonlinear networked systems under unmeasurable premise variables. IEEE Trans Fuzzy Syst 24(5):1233–1245
25. Chang XH, Yang GH (2010) Relaxed stabilization conditions for continuous-time Takagi-Sugeno fuzzy control systems. Inform Sci 180(17):3273–3287
26. Nguang SK, Assawinchaichote W (2003) H_{∞} filtering for fuzzy dynamical systems with D stability constraints. IEEE Trans Circ Syst I: Fundam Theory Appl 50(11):1503–1508
27. Wang XL, Yang GH (2019) Observer-based fault detection for T-S fuzzy systems subject to measurement outliers. Neurocomputing 335:21–36

Chapter 8
Conclusions and Prospects

Since the T-S fuzzy model is proposed by Takagi and Sugeno, research on the one has achieved fruitful results, which have been successfully applied in practical engineering fields. In addition, uncertainties are widely present in dynamic system models, and external unknown factors inevitably act on actual engineering systems, as well as actuators, sensors, and system components suffering from failures without warning, the above mentioned events will lead to the deterioration of system performance, and even cause instability of the system, resulting in unexpected economic losses and casualties. Robust filtering and fault diagnosis provide an effective way to improve the reliability and security of control systems, and related research has attracted much attention. However, it should be pointed out that although many achievements have been made in research on the above-mentioned issues, there are still many problems that need to be further addressed. On the basis of the existing research on T-S fuzzy theory, robust filtering theory, and fault diagnosis theory, the robust filtering and fault diagnosis problems of a class of T-S fuzzy systems are studied in this book. The main contributions are summarized as follows:

1. Fuzzy filters and piecewise fuzzy filters are designed for T-S fuzzy systems with unknown system states and the ones with multiple time-varying delays, respectively, and the corresponding filtering schemes are also proposed to achieve ideal results. Compared with the existing results, the proposed methods can achieve better filtering performance and overcome the shortcomings of the existing ones. In addition, a new event-triggered filtering scheme is proposed for T-S fuzzy systems with bounded disturbances, which realizes that the designed observer gains in the absence of event-triggered mechanisms are also applicable to the case with event-triggered mechanisms.
2. For T-S fuzzy systems in network environments, the introduction of event-triggered mechanisms will lead to the mismatched of premise variables. The improved matching membership function and the online asynchronous premise reconstruction method are established respectively to solve the asynchronous problems of premise variables. At the same time, the fault detection schemes

X.-L. Wang et al., *Robust Filtering and Fault Detection for T-S Fuzzy Systems*,
https://doi.org/10.1007/978-981-96-5818-3_8

are, respectively, constructed to achieve the purpose of detecting faults. Compared with the existing results, the proposed methods overcome the limitations and defects of the existing ones, and achieve better fault detection results.

3. A fault detection observer with adaptive saturation of output errors is designed for T-S fuzzy systems with measurement outliers, and a fault detection scheme is also constructed to achieve the purpose of detecting faults while avoiding false alarms. Compared with the existing results, the proposed fault detection scheme can effectively avoid the occurrence of false alarms.

The problems of robust filtering and fault detection for a class of T-S fuzzy systems are explored in this book, and several difficult problems have also been solved by constructing some new methods. At the same time, it should be recognized that there are still many issues that need to be studied, and future research can be carried out in the following aspects:

1. Asynchronous premise variables for T-S fuzzy systems: The asynchronous problem of premise variables is a common one in research on T-S fuzzy systems. Currently, there are three effective solutions, all of which have the fixed application scenarios, and therefore it is of great theoretical significance to explore a general and effective research method.
2. Incipient fault detection for T-S fuzzy systems: Incipient fault is characterized by small amplitude and low frequency, which makes it difficult to detect. The research on incipient faults has been carried out and achieved some results, but there is still a lack of systematic theory, and better fault detection schemes need to be explored and developed.
3. Fault detection for T-S fuzzy interconnected large-scale systems: Practical engineering systems are usually modeled as fuzzy interconnected large-scale systems, which are composed of multiple subsystems coupled with each other, and the coupling between subsystems will affect the local and overall performance, leading to abnormal difficulty in fault detection. Therefore, the research on fault detection methods for fuzzy interconnected large-scale systems has important theoretical value and practical significance.

In summary, although the research on robust filtering and fault detection for T-S fuzzy systems has been carried out for decades and has achieved fruitful results, there are still many problems that need to be further studied and explored. At the same time, how to extend complex theoretical methods to practical engineering scenarios and solve fundamental problems is a research direction that requires more effort.

MIX
Papier aus verantwortungsvollen Quellen
Paper from responsible sources
FSC® C105338

If you have any concerns about our products,
you can contact us on
ProductSafety@springernature.com

In case Publisher is established outside the EU,
the EU authorized representative is:
Springer Nature Customer Service Center GmbH
Europaplatz 3, 69115 Heidelberg, Germany

Printed by Libri Plureos GmbH
in Hamburg, Germany